Questions and Answers for Electricians Examinations

Revised by Paul Rosenberg

An Audel® Book

ELEVENTH EDITION

Library of Congress Cataloging-in-Publication Data

Questions and answers for electricians examinations / revised by Paul Rosenberg. — 11th ed.
    p.   cm.
    Rev. ed. of: Questions and answers for electricians examinations / Roland E. Palmquist. 10th ed. c1990.
    "An Audel book."
    Includes index.
    ISBN 0-02-077762-0
    1. Electric engineering—Examinations, questions, etc. 2. Electricians—Licenses—United States. I. Rosenberg, Paul. II. Palmquist, Roland E. Questions and answers for electricians examinations.
TK169.Q47   1993
621.319'24'021873—dc20
                          92-33277
                          CIP

10 9 8 7 6 5 4 3 2 1

Contents

Preface

This book is offered as a guide for anyone preparing for the various electricians' license examinations—Apprentice, Journeyman, or Master. These examinations are given periodically by local licensing authorities—either municipal, county, state—or any other agency having legal jurisdiction over the licensing and inspection of work done by electricians.

An electrician is required by law in most localities to secure a license from the enforcing authority in his or her area. This book supplies the license applicant with the required theoretical knowledge and a thorough understanding of the definitions, specifications, and regulations of the *National Electrical Code®*°, on which the majority of these examinations are based.

The *National Electrical Code* has been prepared by the National Fire Protection Association with the assistance of volunteers in all phases of the electrical industry. Anyone in the field of electricity may contribute to the *Code* after first proving the need for change, thereby making this document both flexible and democratic.

Numerous examples in the form of questions and answers are presented throughout this book, thereby enabling the license applicant to gain a complete knowledge of the types of questions asked in an electricians' examination. The tremendous importance of careful study in order to master the fundamental principles underlying each question and answer should be thoroughly impressed on all candidates for licenses in the various grades. Through this process the license applicant can prepare to solve any new or similar problem on an examination.

As with the tenth edition, this eleventh edition contains sections on business competency and test-taking, which will be included in all future editions also.

This eleventh edition will expand its coverage of the newer high-tech installations in a new chapter. The topics covered include fiber optic cables, photovoltaic systems, communications, and the like. The most recent surveys indicate that such installations now account for almost one third of all installations performed by electrical firms. Because of this, more high-tech questions have begun showing up on electrical exams. (We haven't seen too many yet, but it is only a question of time before they arrive in larger numbers.) You can expect that more and more attention will be paid to these areas as time goes on. And, if you have not begun to expand your skills in these areas, you should get started now; high tech is the future of this trade.

The interpretations of the 1993 *NEC* in this book are not necessarily the official observations of the *National Electrical Code* Committee.

Introduction

Tips on Taking Tests

It is the author's experience that for most electricians, knowing how to take a test is almost as important as knowing the technical information, as far as obtaining a passing grade is concerned. A great number of electricians fear tests more than they fear 480 volts.

Really, there is no good reason why this should be so. After all, if hundreds of thousands of men and women can pass these tests, anyone interested who gives a real effort and pays particular attention to some basic rules can succeed. Some basic rules for taking tests are these:

1. Know the material being covered.
2. Know the format of the test.
3. Be physically and mentally prepared on the exam day.
4. RELAX!
5. Work the test the smartest way you can.

The first point, *knowing the material being covered,* is a mandatory prerequisite. Most test failures come from violating this rule. No, it isn't always easy to learn all the material on a test. It requires hours, sometimes many hours, studying, when you'd rather be doing other things. It means that you have to make your brain work harder than it wants to, going over the material again and again. Sorry, but unless you have an exceptional aptitude for learning, there are no shortcuts for hard, intense study. A good study guide (like this book) is about as much help as you can get.

The second rule for taking tests is that you need to *know the format of the test.* Some of the things you need to know are:

How many questions are on the test?
How many questions are open book?
How many are closed book?
Do all questions count for the same number of points?
Is there a penalty for wrong answers?
How much time is allowed for each section of the test?
Who wrote the test?
How will the test be graded?

By knowing the answers to these questions, you can plan your efforts intelligently. For example, if certain questions will count for more points than others, you should be ready to spend more time and effort on those questions. By knowing the time limits, you can calculate how much time you have for each question, etc. Get answers to all of these questions, and consider all of these facts as you prepare for the exam.

Now, as for *being physically and mentally prepared*: I think most readers will be familiar with the way athletes prepare for an event. They make sure they eat the right kinds of food, so that they have enough energy. They get plenty of sleep, and they come to the event planning on winning. The same thing should be done in preparation for a test.

The most important factor is what we mentioned above—planning on winning. Psychologists have found that the results one achieves are directly controlled by what one expects to receive. If you believe that you will do well, you are quite likely to do well; if you believe that you will do poorly, you probably will. Remember, it does not matter what you wish for; what matters is what you actually believe will happen.

If you want to improve your confidence (belief) in your test-taking abilities, picture yourself as having aced the test; refuse to imagine yourself failing; spend as much time studying as is necessary for you to believe in yourself.

On the day of a test, you want to walk in well rested (but not still groggy), having been well fed (but not full), and with a subdued confidence. Generally, heavy studying the night before the test is not a good idea. Do a light review, and leisurely go over a difficult part of the information if you like, but the night before is not the time to get intense. You should have been intense two weeks ago. The night before the test is a time to eat well, and go to bed early. Try not to eat within two or three hours of the test, as it tends to bog you down. It has been said that mental efficiency is highest on an empty stomach. If you need an energy boost right before the exam, try some chocolate milk; it has plenty of food energy, and goes through the digestive system very easily. Confidence is built on a good knowledge of the material to be covered, and the ability to pass with style.

Upon entering the test location, *relaxing* is very important. If you "choke up" during the test, you are automatically taking 5 points off of your score, possibly more. You should have the same attitude as the runner who shows up for a race he knows he will win. He is ready to run his fastest; but he is not nervous, because he knows that his fastest is good enough to win.

Before taking the test, clear your mind, don't get involved with trivial conversations, and then, when it is time to answer the questions, dig into the test full strength.

During the test, first answer all of the easy questions. Do all of the questions that you know the answers to; pass up the hard questions now, and do the ones you know for sure. Then, once you have answered these questions, don't go over them again, and move on to the next group of questions. Now do the questions that will require some work; but do not do the most difficult questions, save them for last. It is silly to waste half of your time on one difficult question. Do the 47 easier questions, and then come back to the 3 especially difficult questions.

Work the test in the smartest possible way. Pay attention to time requirements, books allowed during open book tests, etc. For your electrical exam, you should definitely put tabs on your code book. Bring an electronic calculator with you, and some scratch paper (as long as you are allowed to). Rather than buying a set of code book tabs, I recommend that you do your own. Tab the index, and the sections of the code that you most commonly use. I generally put tabs on:

250-94 & 95.	(Sizes of ground wires.)
Table 310-16.	(Wire ampacities.)

Chapter 9   (Conduit fill tables.)
Article 230.   (Services.)
Table 300-5.   (Burial depths.)
Tables 370-6.   (Number of wires in boxes.)
Article 430.   (Motors.)
Article 450.   (Transformers.)
Article 500.   (Hazardous location wiring.)
Article 700.   (Emergency systems.)

Remember: If hundreds of thousands of other people have passed these tests, you can too—if you prepare.

Business Competency Examinations

In recent years, many municipalities have added business competency examinations to their standard electrical examinations. In reality, they didn't have much choice. Since 1980, the number of licensed electrical contractors has skyrocketed and caused a great number of problems. Most of these problems were not the result of a lack of technical knowledge, but of bad business practices. After some study, the various State Departments of Professional Regulation found out that while the newly-licensed electrical contractors were proficient at trade skills, they were woefully inadequate in business skills.

In an effort to assure that newly-licensed contractors will be knowledgeable in business, new sections have been added to many competency examinations. Typically, 25% of a Master Electrician exam is dedicated to business skills and knowledge. The topics covered are usually:

1. Taxes.
2. Unemployment and worker's compensation.
3. OSHA and safety.
4. Lien laws.
5. Business skills.

To help familiarize you with the various requirements and reference sources, we will briefly discuss each, and then follow with questions and answers.

The *Taxes* section of such tests covers withholding of employee taxes. The information needed to answer these questions can be found in the I.R.S. publication "Circular E." Circular E gives the rules for withholding federal income tax, social security, federal unemployment, and other taxes. These rules are critical, not only for your test, but also in order to operate a business. Many young contractors have gone bankrupt and gotten in legal trouble because they did not know the rules of Circular E. It is critical to study and understand Circular E.

Unemployment compensation is paid directly to the state by the employer. It is not deducted from the employee's wages. Rates vary, and there are a number of requirements for anyone receiving this compensation. All of the required information can be found in a booklet called "Unemployment Compensation Handbook," which is available through various sources, including your Public

Library. As with Circular E, this book is extremely important, and you should thoroughly understand it.

Worker's compensation is handled on the State level, and the requirements vary from state to state; because of this, you will have to get local requirements from your state government. The people who administer your local test should be able to guide you to the right place.

OSHA (Occupational Safety and Health Administration) establishes rules to assure that no employee is subjected to dangers to his or her safety or health. The OSHA regulations can be found in "OSHA Standard 2207, Part 1926." There are too many regulations to memorize but one must be familiar enough with the book to be able to find the answer to any question easily.

Each state has its own *lien laws*. Copies of the regulations must be obtained through your own state government, although the testing agency administering your test can probably tell you exactly how to get them. Liens are very important in the construction business, and have been developed primarily for the benefit of the contractor.

The *business skills* part of the test deals mostly with banks, financing, and basic management skills. As a reference source for the exam, "The Handbook of Small Business Finance" is recommended. This book is published by the U.S. Government. There are many, many other business books available (and we would hope that anyone going into business would read several), but this handbook addresses the material in the test more directly.

You should remember, however, that the business skills covered by these tests are not enough to ensure success in business. In addition to these skills, you will need skill in dealing with people, the ability to analyze a market, the ability to make and follow through on decisions. This test covers only academic business skills; to actually make money, you will need other skills also.

1.   **If a certain employee spends less than half of his time during a pay period performing services that are subject to taxation, how much of his or her pay is taxable?**
   None.

2.   **If an employer fails to make federal income tax deposits when they are due (without a reasonable cause), how large a penalty will he or she be assessed?**
   10%.

3.   **A self-employed person is considered an employee. True or false?**
   False.

4.   **What form must be used to correct errors in withholding taxes?**
   941C.

5.   What would be the take home pay of a worker who claims one deduction, is married, and who earns $205.00 per week (without state or local taxes)?

$173.60:

$205.00
−16.00   Income tax
−15.40   Social Security
173.60

6.   For unemployment taxation, the term employer includes any person or organization that paid _____ or more in wages in any quarter, or had employees at any time in 20 weeks of the year.
$1,500.00

7.   If an employee is paid $325.00 per week, how much social security tax should be deducted from his or her wages?
$24.40: 325.00 x .0765 = $24.40

8.   What form is used to get an employer identification number?
SS-4.

9.   What form must a new employee sign before beginning work?
W-4.

10.   If an employer has _____ or more employees in 20 or more weeks, the employer must file a Form 940 Federal Unemployment Act.
1.

11.   What percentage of his or her wages must an employee contribute for unemployment compensation?
0%, employees do not contribute to unemployment compensation.

12.   *Casual labor* is labor that is occasional, incidental, and not exceeding _____ working days in duration.
10.

13.   Does a temporary light fixture with a reflector that deeply recesses its bulb require a guard?
No.

14.   How long may double cleat ladders be?
15 feet.

15.   The *angle of repose* for average soil is _____ .
45 degrees.

16.   How many gallons of flammable liquid can be stored in a room outside of an approved storage cabinet?
25.

17.   When an interior-hung scaffold is suspended from the beams of a ceiling, what percentage of the rated load must the suspending wire be capable of supporting?
600% (6 times the rated load).

18. What is the proper maintenance procedure for an "ABC" dry chemical stored pressure fire extinguisher?
Check the pressure gauge and the condition of the chemical annually.

19. Workers should not be exposed to impulsive or impact noises louder than _____ decibels.
140.

20. The standard height for a guardrail is _____ .
42 inches.

21. What is the minimum size (OSHA requirement) of a conductor to a ground rod?
#2 AWG copper.

22. For a scaffold with a working load of 75 pounds per square feet, what is the maximum span for a 2" x 12" plank?
7 feet.

23. Loaded powder-activated tools may not be left _____.
Unattended.

24. What should be the predominant color of caution signs?
Yellow.

25. For 225 employees on a construction site, how many toilets must be provided?
5 toilet seats and 5 urinals.

26. When safety belts are used, the maximum distance of fall must be _____.
6 feet.

27. What is the minimum lighting level in a field construction office?
30 footcandles.

28. Manually-handled lumber cannot be stacked higher than _____.
16 feet.

29. What is the term for the claims of a creditor against the assets of a business?
Liabilities.

30. Small Business Administration loans can be guaranteed up to _____.
90%.

31. Would taxes be considered a liability?
Yes.

32. The assets of a business, minus its liabilities, is called its _____.
Equity.

33. Accounts receivable financing is normally based on receivables how old?
70 to 90 days.

34. If your company has gross sales of $210,000.00, and expenses of $198,500.00, what percentage of profit did it make?

5.5%.

35. An agreement by which you get exclusive use of a certain item for a stated period of time is called a _____.

Lease.

36. What is the minimum rate of return sought by most venture capitalists?

It is currently around 11%, but this figure can be modified by changes in interest rates, inflation, etc.

37. A one-year line of credit refers to a note that must be renewed in how many days?

90.

38. What is the term for a plan of cash receipts and expenditures for a certain period of time?

A cash budget.

39. What is the term for the money required to carry accounts receivable, cover payrolls, and to buy products?

Working capital.

40. Is an 18-year-old boy, employed by his parents, exempt from social security tax?

Yes.

41. Is a wife, employed by her husband, subject to social security tax?

No.

Electrical Symbols

To avoid confusion, ASA policy requires that the same symbol not be included in more than one Standard. If the same symbol were to be used in two or more Standards, and one of these Standards were revised, changing the meaning of the symbol, considerable confusion could arise over which symbol was correct, the revised or unrevised.

The symbols in this category include, but are not limited to, those listed below. The reference numbers are the American Standard Y32.2 item numbers.

(MOT)	46.3	*Electric motor*
(GEN)	46.2	*Electric generator*
〰〰	86.1	*Power transformer*
——◁——	82.1	*Pothead (cable termination)*
(WH)	48	*Electric watthour meter*
[CB]	12.2	*Circuit element, e.g., circuit breaker*
⌐	11.1	*Circuit breaker*
⌇	36	*Fusible element*
⟋	76.3	*Single-throw knife switch*

76.2 *Double-throw knife switch*

13.1 *Ground*

7 *Battery*

List of Symbols

1.0 Lighting Outlets

Ceiling　　　**Wall**

1.1 *Surface or pendant incandescent, mercury vapor, or similar lamp fixture*

1.2 *Recessed incandescent, mercury vapor, or similar lamp fixture*

1.3 *Surface or pendant individual fluorescent fixture*

1.4 *Recessed individual fluorescent fixture*

1.5 *Surface or pendant continuous-row fluorescent fixture*

1.6 *Recessed continuous-row fluorescent fixture**

1.7 *Bare-lamp fluorescent strip***

---

*In the case of combination continuous-row fluorescent and incandescent spotlights, use combinations of the above Standard symbols.
**In the case of a continuous-row bare-lamp fluorescent strip above an area-wide diffusion means, show each fixture run, using the Standard symbol; indicate area of diffusing means and type of light shading and/or drawing notation.

(X) $\quad$ $-\text{(X)}$ $\quad$ **1.8** $\quad$ *Surface or pendant exit light*

(XR) $\quad$ $-\text{(XR)}$ $\quad$ **1.9** $\quad$ *Recessed exit light*

(B) $\quad$ $-\text{(B)}$ $\quad$ **1.10** $\quad$ *Blanked outlet*

(J) $\quad$ $-\text{(J)}$ $\quad$ **1.11** $\quad$ *Junction box*

(L) $\quad$ $-\text{(L)}$ $\quad$ **1.12** $\quad$ *Outlet controlled by low-voltage switching when relay is installed in outlet box*

2.0 Receptacle Outlets

Unless noted to the contrary, it should be assumed that every receptacle will be grounded, and will have a separate grounding contact.

Use the uppercase subscript letters described under Section 2 item a-2 of this Standard when weatherproof, explosionproof, or some other specific type of device will be required.

$-\ominus$ $\qquad$ **2.1** $\quad$ **Single receptacle outlet**

$=\ominus$ $\qquad$ **2.2** $\quad$ **Duplex receptacle outlet**

$\equiv\oplus$ $\qquad$ **2.3** $\quad$ **Triplex receptacle outlet**

⊕	**2.4**	*Quadruplex receptacle outlet*
⊖	**2.5**	*Duplex receptacle outlet—split wired*
⊕	**2.6**	*Triplex receptacle outlet—split wired*
△*	**2.7**	*Single special-purpose receptacle outlet°*
△*	**2.8**	*Duplex special-purpose receptacle outlet°*
⊕R	**2.9**	*Range outlet*
▲DW	**2.10**	*Special-purpose connection or provision for connection. Use subscript letters to indicate function (DW—dishwasher; CD—clothes dryer, etc.)*
⊕ X"	**2.11**	*Multioutlet assembly. Extend arrows to limit of installation. Use appropriate symbol to indicate type of outlet. Also indicate spacing of outlets as x inches.*
Ⓒ	**2.12**	*Clock Hanger Receptacle*
Ⓕ	**2.13**	*Fan Hanger Receptacle*

°Use numeral or letter, either within the symbol or as a subscript alongside the symbol keyed to explanation in the drawing list of symbols, to indicate type of receptacle or usage.

2.14 Floor Single Receptacle Outlet

2.15 Floor Duplex Receptacle Outlet

2.16 Floor Special-Purpose Outlet°

2.17 Floor Telephone Outlet—Public

2.18 Floor Telephone Outlet—Private

Not a part of the Standard: example of the use of several floor outlet symbols to identify a 2-, 3-, or more-gang floor outlet

2.19 Underfloor Duct and Junction Box for Triple, Double or Single Duct System as indicated by the number of parallel lines

Not a part of the Standard: example of use of various symbols to identify location of different types of outlets or connections for underfloor duct or cellular floor systems

2.20 Cellular Floor Header Duct

°Use numeral or letter, either within the symbol or as a subscript alongside the symbol keyed to explanation in the drawing list of symbols, to indicate type of receptacle or usage.

3.0 Switch Outlets

S	*3.1*	*Single-pole switch*
S₂	*3.2*	*Double-pole switch*
S₃	*3.3*	*Three-way switch*
S₄	*3.4*	*Four-way switch*
S_K	*3.5*	*Key-operated switch*
S_P	*3.6*	*Switch and pilot lamp*
S_L	*3.7*	*Switch for low-voltage switching system*
S_LM	*3.8*	*Master switch for low-voltage switching system*
⊖S	*3.9*	*Switch and single receptacle*
⊜S	*3.10*	*Switch and double receptacle*
S_D	*3.11*	*Door switch*
S_T	*3.12*	*Time switch*
S_CB	*3.13*	*Circuit-breaker switch*
S_MC	*3.14*	*Momentary contact switch or push-button for other than signaling system*

Signaling System Outlets

4.0 Institutional, Commercial, and Industrial Occupancies

These symbols are recommended by the American Standards Association, but are not used universally. The reader should remember not to assume that these symbols will be used on any certain plan, but to always check the symbol list on the plans, and verify if these symbols are actually used.

Basic Symbol

Examples of Individual Item Identification (Not a part of the Standard)

4.1 I. **Nurse Call System Devices (any type)**

Nurses' Annunciator (can add a number after it as ─○ 24 to indicate number of lamps)

Call station, single cord, pilot light

Call station, double cord, microphone speaker

Corridor dome light, 1 lamp

Transformer

Any other item on same system—use numbers as required.

4.2 II. **Paging System Devices (any type)**

Keyboard

Flush annunciator

2-face annunciator

Any other item on same system—use numbers as required.

Basic Symbol	Examples of Individual Item Identification (Not a part of the Standard)	

+—□

4.3 III. Fire Alarm System Devices (any type) including Smoke and Sprinkler Alarm Devices

+—[1] *Control panel*

+—[2] *Station*

+—[3] *10" Gong*

+—[4] *Pre-signal chime*

+—[5] *Any other item on same system—use numbers as required.*

+—◇

4.4 IV. Staff Register System Devices (any type)

+—◈① *Phone operators' register*

+—◈② *Entrance register—flush*

+—◈③ *Staff room register*

+—◈④ *Transformer*

+—◈⑤ *Any other item on same system—use numbers as required.*

Basic Symbol	Examples of Individual Item Identification (Not a part of the Standard)		

4.5 V. Electric Clock System Devices (any type)

Master clock

12" Secondary—flush

12" Double dial—wall-mounted

18" Skeleton dial

Any other item on same system—use numbers as required.

4.6 VI. Public Telephone System Devices

Switchboard

Desk phone

Any other item on same system—use numbers as required.

4.7 VII. Private Telephone System Devices (any type)

Switchboard

Wall phone

Any other item on same system—use numbers as required.

Basic Symbol	Examples of Individual Item Identification (Not a part of the Standard)		

| | | 4.8 | VIII. **Watchman System Devices (any type)** |

| | 1 | | *Central station* |

| | 2 | | *Key station* |

| | 3 | | *Any other item on same system—use numbers as required.* |

| | | 4.9 | IX. **Sound System** |

| | 1 | | *Amplifier* |

| | 2 | | *Microphone* |

| | 3 | | *Interior speaker* |

| | 4 | | *Exterior speaker* |

| | 5 | | *Any other item on same system— use numbers as required.* |

| | | 4.10 | X. **Other Signal System Devices** |

| | 1 | | *Buzzer* |

| | 2 | | *Bell* |

| | 3 | | *Pushbutton* |

| | 4 | | *Annunciator* |

| | 5 | | *Any other item on same system— use numbers as required.* |

Signaling System Outlets

5.0 Residential Occupancies

When a descriptive symbol list is not employed, use the following signaling system symbols to identify standardized, residential-type, signal-system items on residential drawings. Use the basic symbols with a descriptive symbol list when other signal-system items are to be identified.

	5.1	**Pushbutton**
	5.2	**Buzzer**
	5.3	**Bell**
	5.4	**Combination bell-buzzer**
CH	5.5	**Chime**
	5.6	**Annunciator**
D	5.7	**Electric door opener**
M	5.8	**Maid's signal plug**
	5.9	**Interconnection box**
BT	5.10	**Bell-ringing transformer**
	5.11	**Outside telephone**
	5.12	**Interconnecting telephone**
R	5.13	**Radio outlet**
TV	5.14	**Television outlet**

6.0 Panelboards, Switchboards, and Related Equipment

6.1 *Flush-mounted panelboard and cabinet**

6.2 *Surface-mounted panelboard and cabinet**

6.3 *Switchboard, power control center, unit substations*— should be drawn to scale*

6.4 *Flush-mounted terminal cabinet.* In small-scale drawings the TC may be indicated alongside the symbol.*

6.5 *Surface-mounted terminal cabinet.* In small-scale drawings the TC may be indicated alongside the symbol.*

6.6 *Pull box (identify in relation to wiring section and sizes)*

6.7 *Motor or other power controller**

6.8 *Externally-operated disconnection switch**

6.9 *Combination controller and disconnection means**

---

*Identify by notation or schedule.

7.0 Bus Ducts and Wireways

`⊏T⊐ ⊏T⊐ ⊏T⊐`	**7.1 Trolley duct***
`⊏B⊐ ⊏B⊐ ⊏B⊐`	**7.2 Busway (service, feeder, or plug-in)***
`⊏C⊐ ⊏C⊐ ⊏C⊐`	**7.3 Cable trough ladder or channel***
`⊏W⊐ ⊏W⊐ ⊏W⊐`	**7.4 Wireway***

8.0 Remote Control Stations for Motors or other Equipment*

	8.1 Pushbutton station
`F`	**8.2 Float switch—mechanical**
`L`	**8.3 Limit switch—mechanical**
`P`	**8.4 Pneumatic switch—mechanical**
	8.5 Electric eye—beam source
	8.6 Electric eye—relay
	8.7 Thermostat

---

*Identify by notation or schedule.

9.0 Circuiting

Wiring method identification by notation on drawing or in specification.

———————————————   **9.1 Wiring concealed in ceiling or wall**

— — — — — —   **9.2 Wiring concealed in floor**

-----------------   **9.3 Wiring exposed**

Note: Use heavyweight line to identify service and feeders. Indicate empty conduit by notation CO (conduit only).

2   1
———————————→   **9.4 Branch-circuit home run to panelboard. Number of arrows indicates number of circuits. (A numeral at each arrow may be used to identify circuit number.) Note: Any circuit without further identification indicates two-wire circuit. For a greater number of wires, indicate with cross lines, e.g.:**

3 wires;
—————/////—————

4 wires, etc.
—————/////—————   *Unless indicated otherwise, the wire size of the circuit is the minimum size required by the specification.*

Identify different functions of wiring system, e.g., signaling system by notation or other means.

———————————o   **9.5 Wiring turned up**

———————————•   **9.6 Wiring turned down**

10.0 Electric Distribution or Lighting System, Underground

M
          10.1 Manhole*

H
          10.2 Handhole*

TM
          10.3 Transformer manhole or vault*

TP
          10.4 Transformer pad*

10.5 Underground direct burial cable. Indicate type, size, and number of conductors by notation or schedule.

10.6 Underground duct line. Indicate type, size, and number of ducts by cross-section identification of each run by notation or schedule. Indicate type, size, and number of conductors by notation or schedule.

10.7 Streetlight standard feed from underground circuit*

*Identify by notation or schedule.

11.0 Electric Distribution or Lighting System Aerial

11.1 Pole°

11.2 Streetlight and bracket°

11.3 Transformer°

11.4 Primary circuit°

11.5 Secondary circuit°

11.6 Down guy

11.7 Head guy

11.8 Sidewalk guy

11.9 Service weather head°

4 Arrester, Lighting Arrester (Electric surge, etc.) Gap

4.1 General

4.2 Carbon block

Block, telephone protector

The sides of the rectangle are to be approximately in the ratio of 1 to 2, and the space between rectangles shall be approximately equal to the width of a rectangle.

°Identify by notation or schedule.

4.3 *Electrolytic or aluminum cell*
This symbol is not composed of arrowheads.

4.4 *Horn gap*

4.5 *Protective gap*
These triangles shall not be filled.

4.6 *Sphere gap*

4.7 *Valve or film element*

4.8 *Multigap, general*

4.9 *Application: gap plus valve plus ground, 2-pole*

7 Battery

The long line is always positive, but polarity may be indicated in addition.
Example:

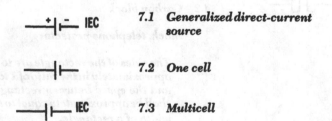

7.1 *Generalized direct-current source*

7.2 *One cell*

7.3 *Multicell*

7.3.1 *Multicell battery with 3 taps*

7.3.2 *Multicell battery with adjustable tap*

11 Circuit Breakers

If it is desired to show the condition causing the breaker to trip, the relay-protective-function symbols in item 66.6 may be used alongside the break symbol.

IEC

11.1 *General*

IEC

11.2 *Air circuit breaker, if distinction is needed; for alternating-current breakers rated at 1,500 volts or less and for all direct-current circuit breakers.*

11.2.1 *Network protector*

See Note 11.3A

11.3 *Circuit breaker, other than covered by item 11.2. The symbol in the right column is for a 3-pole breaker.*

See Note 11.3A

11.3.1 *On a connection or wiring diagram, a 3-pole single-throw circuit breaker (with terminals shown) may be drawn as shown.*

Note 11.3A—On a power diagram, the symbol may be used without other identification. On a composite drawing where confusion with the general circuit element symbol (item 12) may result, add the identifying letters CB inside or adjacent to the square.

11.4   Applications

11.4.1   3-pole circuit breaker with thermal overload device in all 3 poles

11.4.2   3-pole circuit breaker with magnetic overload device in all 3 poles

11.4.3   3-pole circuit breaker, drawout type

13 Circuit Return

13.1 Ground

(A)   A direct conducting connection to the earth or body of water that is a part thereof

(B)   A conducting connection to a structure that serves a function similar to that of an earth ground (that is, a structure such as a frame of an air, space, or land vehicle that is not conductively connected to earth)

13.2 Chassis or frame connection
A conducting connection to a chassis or frame of a unit. The chassis or frame may be at a substantial potential with respect to the earth or structure in which this chassis or frame is mounted.

13.3 Common connections
Conducting connections made to one another. All like-designated points are connected. *The asterisk is not a part of the symbol. Identifying valves, letters, numbers, or marks shall replace the asterisk.*

15 Coil, Magnetic Blowout*

23 Contact, Electrical

For buildups or forms using electrical contacts, see applications under CONNECTOR (item 19), RELAY (item 66), SWITCH (item 76). See DRAFTING PRACTICES (item 0.4.6).

*The broken line — - — indicates where line connection to a symbol is made and is not a part of the symbol.

23.1.1 *Fixed contact for jack, key, relay, etc.*

23.1.2 *Fixed contact for switch*

23.1.3 *Fixed contact for momentary switch*
   See SWITCH (item 76.8 and 76.10).

23.1.4 *Sleeve*

23.2.1 *Adjustable or sliding contact for resistor, inductor, etc.*

23.2.2 *Locking*

23.2.3 *Segment; bridging contact*

23.2.4 *Nonlocking*
   See SWITCH (items 76.12.3 and 76.12.4).

23.2.5 *Vibrator reed*

23.2.6 *Vibrator split reed*

23.2.7 *Rotating contact (slip ring) and brush*

It is standard procedure to show a contact by a symbol which indicates the circuit condition produced when the actuating device is in the nonoperated, or deenergized, position. It may be necessary to add a clarifying note explaining the proper point at which the contact functions—the point where the actuating device (mechanical, electrical, etc.) opens or closes due to changes in pressure, level, flow, voltage, current, etc. When it is necessary to show contacts in the oper-

ated, or energized, condition—and where confusion would otherside result—a clarifying note shall be added to the drawing. Contacts for circuit breakers or auxiliary switches, etc., may be designated as shown below:

(a)   Closed when device is in energized or operated position

(b)   Closed when device is in deenergized or nonoperated position

(aa) Closed when operating mechanism of main device is in energized or operated position

(bb) Closed when operating mechanism of main device is in deenergized or nonoperated position

[See American Standard C37.2-1962 for details.]

In the parallel-line contact, symbols showing the length of the parallel lines shall be approximately 1¼ times the width of the gap (except for item 23.6)

23.3.1 Closed contact (break)
        See also SWITCHING FUNC-
        TION (item 77).

23.3.2 Open contact (make)
        See also SWITCHING FUNC-
        TION (item 77).

23.3.3 *Transfer*
See also SWITCHING
FUNCTION (item 77).

23.3.4 *Make-before-break*

23.4 *Application: open contact with time closing (TC) or time delay closing (TDC) feature*

23.5 *Application: closed contact with time opening (TO) or time delay opening (TDO) feature*

23.6 *Time sequential closing*

24 Contactor

See also RELAY (item 66).

Contactor symbols are derived from fundamental contact, coil, and mechanical connection symbols, and should be employed to show contactors on complete diagrams. A complete diagram of the actual contactor device is constructed by combining the abovementioned fundamental symbols for mechanical connections, control circuits, etc.

Mechanical interlocking should be indicated by notes.

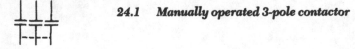

24.1 *Manually operated 3-pole contactor*

24.2 *Electrically operated 1-pole contact or with series blowout coil*
See Note 24.2A.

24.3 *Electrically operated 3-pole contactor with series blowout coils; 2 open and 1 closed auxiliary contacts (shown smaller than the main contacts)*

24.4 *Electrically operated 1-pole contactor with shunt blowout coil*

46 Machine, Rotating

46.1 *Basic*

46.2 *Generator (general)*

46.3 *Motor (general)*

46.4 *Motor, multispeed*

USE BASIC MOTOR SYMBOL AND NOTE SPEEDS

46.5 *Rotating armature with commutator and brushes**

Note 24.2A—The asterisk is not a part of the symbol. Always replace the asterisk by a device designation.

*The broken line - - - indicates where line connection to a symbol is made and is not a part of the symbol.

	46.6	Field, generator or motor
		Either symbol of item 42.1 may be used in the following items.
�—ᴍ— IEC	46.6.1	Compensating or commutating
�—ᴍ— IEC	46.6.2	Series
�—ᴍᴍ— IEC	46.6.3	Shunt, or separately excited
	46.6.4	Magnet, permanent
		See item 47.
	46.7	Winding symbols
		Motor and generator winding symbols may be shown in the basic circle using the following representation.
⊕	46.7.1	1-phase
⊗	46.7.2	2-phase
⦻	46.7.3	3-phase wye (ungrounded)
⦻⊸	46.7.4	3-phase wye (grounded)
◬	46.7.5	3-phase delta
✴	46.7.6	6-phase diametrical
◉	46.7.7	6-phase double-delta
	46.8	Direct-current machines; applications

46.8.1 *Separately excited direct-current generator or motor**

46.8.2 *Separately excited direct-current generator or motor; with commutating or compensating field winding or both**

46.8.3 *Compositely excited direct-current generator or motor; with commutating or compensating field winding or both**

46.8.4 *Direct-current series motor or 2-wire generator**

46.8.5 *Direct-current series motor or 2-wire generator; with commutating or compensating field winding or both**

46.8.6 *Direct-current shunt motor or 2-wire generator**

46.8.7 *Direct-current shunt motor or 2-wire generator; with commutating or compensating field winding or both**

*The broken line -—- indicates where line connection to a symbol is made and is not a part of the symbol.

46.8.8   *Direct-current, permanent-magnet field generator or motor**

46.8.9   *Direct-current, compound motor or 2-wire generator or stabilized shunt motor**

46.8.10   *Direct-current compound motor or 2-wire generator or stabilized shunt motor; with commutating or compensating field winding or both**

46.8.11   *Direct-current, 3-wire shunt generator**

46.8.12   *Direct-current, 3-wire shunt generator; with commutating or compensating field winding or both**

*The broken line -—- indicates where line connection to a symbol is made and is not a part of the symbol.

46.8.13 *Direct-current, 3-wire compound generator**

46.8.14 *Direct-current, 3-wire compound generator; with commutating or compensating field winding or both**

46.8.15 *Direct-current balancer, shunt wound**

46.9 *Alternating-current machines; application*

---

*The broken line - - - indicates where line connection to a symbol is made and is not a part of the symbol.

46.9.1 *Squirrel-cage induction motor or generator, split-phase induction motor or generator, rotary phase converter or repulsion motor**

46.9.2 *Wound-rotor induction motor, synchronous induction motor, induction generator, or induction frequency converter**

46.9.3 *Alternating-current series motor**

48 Meter Instrument

As indicated in Note 48A, the asterisk is not part of the symbol and should always be replaced with one of the letter combinations listed below, according to the meter's function. This is not necessary if some other identification is provided in the circle and described in the diagram.

A   *Ammeter IEC*

AH   *Ampere-hour*

CMA   *Contact-making (or breaking) ammeter*

CMC   *Contact-making (or breaking) clock*

CMV   *Contact-making (or breaking) voltmeter*

CRO   *Oscilloscope or cathode-ray oscillograph*

DB   *DB (decibel) meter*

DBM   *DBM (decibels referred to 1 milliwatt) meter*

---

*The broken line - — - indicates where line connection to a symbol is made and is not a part of the symbol.

DM	*Demand meter*
DTR	*Demand-totalizing relay*
F	*Frequency meter*
G	*Galvanometer*
GD	*Ground detector*
I	*Indicating*
INT	*Integrating*
μA or	
UA	*Microammeter*
MA	*Milliammeter*
NM	*Noise meter*
OHM	*Ohmmeter*
OP	*Oil pressure*
OSCG	*Oscillograph string*
PH	*Phasemeter*
PI	*Position indicator*
PF	*Power factor*
RD	*Recording demand meter*
REC	*Recording*
RF	*Reaction factor*
SY	*Synchroscope*
TLM	*Telemeter*
T	*Temperature meter*
THC	*Thermal converter*
TT	*Total time*
V	*Voltmeter*
VA	*Volt-ammeter*

VAR	*Varmeter*
VARH	*Varhour meter*
VI	*Volume indicator; meter, audio level*
VU	*Standard volume indicator; meter, audio level*
W	*Wattmeter*
WH	*Watthour meter*

58 Path, Transmission, Conductor, Cable, Wiring

_____ IEC   58.1   *Guided path, general*
The entire group of conductors, or the transmission path required to guide the power or symbol, is shown by a single line. In coaxial and waveguide work, the recognition symbol is employed at the beginning and end of each type of transmission path as well as at intermediate points to clarify a potentially confusing diagram. For waveguide work, the mode may be indicated as well.

_____ IEC   58.2   *Conductive path or conductor; wire*

58.2.1 *Two conductors or conductive paths of wires*

—//—IEC _____IEC

58.2.2 *Three conductors or conductive paths of wires*

—///—IEC =====IEC

58.2.3   *"n" conductors or conductive paths of wires*

58.5   *Crossing of paths or conductors not connected*
       The crossing is not necessarily at a 90-degree angle.

58.6   *Junction of paths or conductors*

58.6.1   *Junction of paths or conductors*

58.6.1.1   *Application: junction of paths, conductor, or cable. If desired, indicate path type or size.*

58.6.1.2   *Application: splice (if desired) of same size cables. Junction of conductors of same size or different size cables. If desired, indicate sizes of conductors.*

58.6.2   *Junction of connected paths, conductors, or wires*

       OR ONLY IF REQUIRED BY SPACE LIMITATION

63 Polarity Symbol

+ **EC**　　　　　*63.1　　Positive*

− **EC**　　　　　*63.2　　Negative*

76 Switch

See also FUSE (item 36); CONTACT, ELECTRIC (item 23); and DRAFTING PRACTICES (items 0.4.6 and 0.4.7).

Switch symbols may be constructed of the fundamental symbols for mechanical connections, contacts, etc.

In standard procedure, a switch is represented in the nonoperating, or deenergized position. In the case of switches that have two or more positions in which no operating force is applied and for those switches (air-pressure, liquid-level, rate-of-flow, etc.) that may be actuated by a mechanical force, the point at which the switch functions should be described in a clarifying note.

In cases where the basic switch symbols (items 76.1–76.4) are used in a diagram in the closed position, the terminals must be included for clarity.

IEC　　*76.1　　Single throw, general*

76.2　　Double throw, general

76.2.1　Application: 2-pole double-throw switch with terminals shown

76.3　　Knife switch, general

76.6   *Pushbutton, momentary or spring return*

76.6.1   *Circuit closing (make)*

76.6.2   *Circuit opening (break)*

76.6.3   *Two-circuit*

76.7   **Pushbutton, maintained or not spring return**

76.7.1   *Two-circuit*

86 Transformer

IEC

86.1   *General*

Either winding symbol may be used. In the following items, the left symbol is used. Additional windings may be shown or indicated by a note. For power transformers use polarity marking H_1, X_1, etc., from American Standard C6.1-1956.

For polarity markings on current and potential transformers, see items 86.16.1 and 86.17.1.

In coaxial and waveguide circuits, this symbol will represent a taper or step transformer without mode change.

86.1.1. *Application: transformer with direct-current connections and mode suppression between two rectangular waveguides*

86.2 *If it is desired especially to distinguish a magnetic-core transformer*

86.2.1 *Application: shielded transformer with magnetic core shown*

86.2.2 *Application: transformer with magnetic core shown and with a shield between windings. The shield is shown connected to the frame.*

86.6 *With taps, 1-phase*

86.7 *Autotransformer, 1-phase*

86.7.1 *Adjustable*

86.13 *1-phase, 2-winding transformer*

86.13.1 *3-phase bank of 1-phase, 2-winding transformer*

See American Standard C6.1-1965 for interconnections for complete symbol.

86.14 *Polyphase transformer*

86.16 *Current transformer(s)*

86.16.1 *Current transformer with polarity marking. Instantaneous direction of current into one polarity mark corresponds to current out of the other polarity mark.*

Symbol used shall not conflict with item 77 when used on same drawing.

86.16.2 *Bushing-type current transformer**

86.17 *Potential transformer(s)*

86.17.1 *Potential transformer with polarity mark. Instantaneous direction of current into one polarity mark corresponds to current out of the other polarity mark.*

Symbol used shall not conflict with item 77 when used on same drawing.

86.18 *Outdoor metering device*

86.19 *Transformer winding connection symbols*

For use adjacent to the symbols for the transformer windings.

IEC

86.19.1 *2-phase 3-wire, grounded*

---

*The broken line - — - indicates where line connection to a symbol is made and is not a part of the symbol.

IEC	*86.19.1.1*	2-phase 3-wire, grounded
	86.19.2	2-phase 4-wire
IEC	*86.19.2.1*	2-phase 5-wire, grounded
IEC	*86.19.3*	3-phase 3-wire, delta or mesh
IEC	*86.19.3.1*	3-phase 3-wire, delta, grounded
	86.19.4	3-phase 4-wire, delta, ungrounded
	86.19.4.1	3-phase 4-wire, delta, grounded
IEC	*86.19.5*	3-phase, open-delta
IEC	*86.19.5.1*	3-phase, open-delta, grounded at common point
	86.19.5.2	3-phase, open-delta, grounded at middle point of one transformer
IEC	*86.19.6*	3-phase, broken-delta

IEC   86.19.7   *3-phase, wye or star, ungrounded*

IEC   86.19.7.1   *3-phase, wye, grounded neutral*

                                The direction of the stroke representing the neutral can be arbitrarily chosen.

IEC   86.19.8   *3-phase 4-wire, ungrounded*

CHAPTER 1

Review Definitions

Definitions are covered in Article 100 of the *NEC (National Electrical Code)*. The questions that follow will not cover all of the definitions but only the more pertinent ones. The answers given here are the author's; refer to Article 100 of the *NEC* for their definitions. Some definitions appear in the *Code* elsewhere than Article 100 (see *NEC*, Index).

1-1 What is "accessible" as applied to wiring methods?
Readily available to inspection, repair, removal, etc., without disturbing the building structure or finish. Not permanently closed by structure or finish of buildings.

1-2 What is "accessible" as applied to equipment?
Equipment that may be readily reached without climbing over obstacles, that is not in locked or other hard-to-get-at areas, e.g., panelboards in kitchen cabinets that are mounted in or on the walls above washers and dryers or in closets or bathrooms. Service-entrance equipment that can only be reached by going into a basement, behind a stairway, or around some other obstacle would most certainly not be termed accessible.

1-3 What does "ampacity" mean?
The amount of flowing current (in amperes) that a conductor can carry continuously for specific use conditions and not exceed the temperature rating of the conductor (see *NEC* Section 310-10).

1-4 What is a branch circuit?
The conductors that originate from the load side of a fuse or circuit breaker and extend on to provide power for the entire circuit in a particular run.

1-5   What is a building?
A building may be a structure that stands by itself, or one that is separated from another by a fire wall.

1-6   What does "dead front" mean?
No live (energized) parts exposed where a person operates electrical equipment.

1-7   What does "approved" mean?
Any appliance, wiring material, or other electrical equipment that is acceptable to the enforcing authorities. Frequently, the Underwriters Laboratory is the most acceptable authority to inspectors. Do not be fooled by a UL label on the motor or cord of an appliance without such a label on the entire article (see *NEC*, Section 110-2).

1-8   What does "identified" mean as it is applied to equipment?
It means that the equipment will be suitable for the particular use or environment, and that it has been evaluated by a qualified electrical testing organization and features a product listing or label indicating its suitability.

1-9   What is a branch circuit?
A branch circuit is the portion of a wiring system that extends beyond the last overcurrent-protective device. In interpreting this, you must not consider the thermal cutout or the motor overload protection as the beginning of the branch circuit. The branch circuit actually begins at the final fusing or circuit-breaker point where the circuit breaks off to supply the motor.

1-10   What is an appliance branch circuit?
This is the circuit supplying one or more outlets connecting appliances only; there is no permanently-connected lighting on this circuit, except the lighting that may be built into the appliance. This term is most often used in connection with Section 220-4(b) and (c) of the *NEC*, which refers to outlets for small appliance loads in kitchens, laundries, pantries, and dining and breakfast rooms of dwellings.

1-11   What is a general purpose circuit?
This is a branch circuit to which lighting and/or appliances may be connected. Lighting may be connected to this circuit, whereas in 1-10 lighting cannot be connected unless the conditions of *NEC* Exception No. 3 in Section 210-52(b)(1) can be met.

1-12   What is a multiwire branch circuit?
A multiwire branch circuit has two or more ungrounded conductors with a potential difference between them and also has a grounded (neutral) conductor with an equal potential difference between it and each of the other wires; e.g., a three-wire 120/240-volt system or a 120/208-volt wye system, using two- or three-phase conductors and a grounded conductor. However, in either case, the "hot" wires must not be tied to one phase but must be connected to different phases to make the system a multiwire circuit (see *NEC*, Section 210-4).

1-13 What is a circuit breaker?

A device that is designed not only to open and close a circuit non-automatically but also to open the circuit automatically at a predetermined current-overload value. The circuit breaker may be thermally or magnetically operated; however, ambient temperatures affect the operation of the thermally operated type, so that the trip value of the current is not as stable as with the magnetic type.

1-14 What is a current-carrying conductor?

A conductor that is expected to carry current under normal operating conditions.

1-15 What is a noncurrent-carrying conductor?

One that carries current only in the event of a malfunction of equipment or wiring. An equipment grounding conductor is a good example; it is employed for protection and is quite a necessary part of the wiring system, but it is not used for carrying normal currents, except in the case of faulty operation where it aids in tripping the overcurrent-protective device.

1-16 What is a pressure connector (solderless)?

A device that establishes a good electrical connection between two or more conductors by some means of mechanical pressure. A pressure connector is used in place of soldering connections and is required to be of an approved type.

1-17 What is meant by "demand factor"?

This is the ratio between the maximum demand on a system or part of a system and the total connected load on the same system or part of the system.

1-18 What is meant by "dust-tight"?

The capacity to keep dust out of the enclosing case so that dust cannot interfere with normal operation. This is discussed further in connection with Articles 500 and 502 of the *NEC*, which both cover hazardous (classified) locations.

1-19 What is meant by "explosionproof apparatus"?

Apparatus enclosed in a case that is capable of sustaining an explosion that may occur within itself and is also capable of preventing ignition of specified gases or vapors surrounding the enclosure by sparks, flashes, or explosion of the gases or vapors within; it must also operate at a temperature that will not ignite any inflammable atmosphere or residue surrounding it. If an explosion does occur within the equipment, the gases are allowed to escape either by a ground joint or by threads, and the escaping gases are thereby cooled to a temperature low enough to inhibit the ignition of any external gases.

1-20 What is a feeder?

The circuit conductors between the service equipment or the source of a separately derived system and the final branch-circuit overcurrent device or devices. Generally, feeders are comparatively large in size and supply a feeder panel, which is composed of a number of branch-circuit overcurrent devices.

1-21 What is a fitting?

A mechanical device, such as a locknut or bushing, that is intended primarily for a mechanical, rather than an electrical, function.

1-22 What is meant by a "ground"?

An electrical connection, either accidental or intentional, that exists between an electrical circuit or equipment and earth, or some other electrical conducting body that serves in place of the earth.

1-23 What does "grounded" mean?

Connected to earth or to some other conducting body which serves in place of the earth.

1-24 What is a grounded conductor?

A system or circuit conductor which is intentionally grounded.

1-25 What is a grounding conductor?

A conductor that is used to connect equipment, devices, or wiring systems with grounding electrodes.

1-26 What is a grounding conductor (equipment)?

It is the conductor used to connect noncurrent-carrying metal parts of equipment, raceways, and other enclosures to the system grounding conductor at the service and/or the grounding electrode conductor.

1-27 What is a grounding electrode conductor?

It is a conductor used to connect the grounding electrode to the equipment grounding conductor and/or to the grounded conductor of the circuit at the service.

1-28 What is a dwelling unit?

A dwelling unit includes one or more rooms used by one or more persons, with space for sleeping, eating, living, and a permanent provision for cooking and sanitation.

1-29 What is an outlet?

A point in the wiring system at which current is taken to supply some equipment.

1-30 What is meant by "raintight"?

Capable of withstanding a beating rain without resulting in the entrance of water.

1-31 What is a receptacle?

A receptacle is a contact device installed at the outlet for the connection of a single attachment plug. A single receptacle is a single device with no other contact device on the same yoke. A multiple receptacle is a single device containing two or more receptacles.

1-32 What is "rainproof"?

So constructed, protected, or treated as to prevent rain from interfering with the successful operation of the apparatus.

Pay particular attention to the following questions; they involve services and are probably among the most misused of any definitions in the *NEC*.

1-33 What is meant by the term "service"?
The conductors and equipment for delivering electrical energy from the secondary distribution system—the street main, the distribution feeder, or the transformer—to the wiring system on the premises. This includes the service-entrance equipment and the grounding electrode.

1-34 What are service conductors?
The portion of the supply conductors that extend from the street main, duct, or transformers to the service-entrance equipment of the premises supplied. For overhead conductors, this includes the conductors from the last line pole (this does not mean the service pole) to the service equipment.

1-35 What is a service cable?
Service conductors that are made up in the form of a cable and are normally referred to as SE cable, or USE cable (see *NEC*, Article 338).

1-36 What is meant by "service drop"?
The overhead conductors from the last pole or other aerial support to and including the splices, if any, connecting to the service-entrance conductors at the building or other structure. If there is a service pole with a meter on it, such as a farm service pole, the service drop does not stop at the service pole; all wires extending from this pole to a building or buildings are service drops, as well as the conductors from the last line pole to the service pole (see *NEC*, Article 100 and Section 230-21).

1-37 What are service-entrance conductors (overhead system)?
That portion of the service conductors between the terminals of service equipment and a point outside the building, clear of building walls, where they are joined by a splice or tap to the service drop, street main, or other source of supply.

1-38 What are service-entrance conductors (underground system)?
The service conductors between the terminals of the service equipment and the point of connection to the service lateral. Where service equipment is located outside the building walls, there may be no service-entrance conductors, or they may be entirely outside the building.

1-39 What are sets of service-entrance conductors?
Sets of service-entrance conductors are taps run from main service conductors to service equipment.

1-40 What is meant by "service equipment"?
The necessary equipment, usually consisting of circuit breakers or switches and fuses and their accessories, located near the point of entrance of supply conductors to a building or other structure, or an otherwise defined area, and intended to constitute the main control and means of cutoff of the supply.

1-41 What is meant by "service lateral"?
The underground service conductors between the street main, including any risers at the pole or other structure or from transformers, and the first point of connection to the service-entrance conductors in a terminal box. The point of connection is considered to be the point of entrance of the service conductors into the building.

1-42 What is meant by "service raceway"?
The rigid metal conduit, electrical metallic tubing, or other raceway that encloses service-entrance conductors.

1-43 What is meant by "special permission"?
The written consent of the authority enforcing the NEC. Under most circumstances, this is a local electrical inspector.

1-44 What is meant by a "general-use switch"?
A device intended for use as a switch in general distribution and branch circuits. It is rated in amperes and is capable of interrupting its rated current at its rated voltage.

1-45 What is meant by a "T-rated switch"?
An AC general-use snap switch that is capable of use on resistive and inductive loads that do not exceed the ampere rating at the voltage involved, on tungsten-filament lighting loads that do not exceed the ampere rating at 120 volts, and on motor loads that do not exceed 80% of their ampere rating at the rated voltage.

1-46 What is meant by an "isolating switch"?
A switch that is intended for isolating an electric circuit from its source of power. It has no interrupting rating and is intended to be operated only after the circuit has been opened by some other means.

1-47 What is meant by a "motor-circuit switch"?
A switch, rated in horsepower, that is capable of interrupting the maximum operating overload current of a motor of the same horsepower rating as the switch at the rated voltage.

1-48 What is meant by "watertight"?
So constructed that moisture will not enter the enclosing case.

1-49 What is meant by "weatherproof"?
So constructed or protected that exposure to the weather will not interfere with successful operation. Raintight or watertight may fulfill the requirements for "weatherproof." However, weather conditions vary, and consideration should be given to the conditions resulting from snow, ice, dust, and temperature extremes.

1-50 What is meant by the "voltage" of a circuit?
This is the greatest effective difference of potential (root-mean-square difference of potential) that exists between any two conductors of a circuit. On various systems, such as 3-phase 4-wire, single-phase 3-wire, and 3-wire direct current, there may be various circuits of numerous voltages.

CHAPTER 2

Ohm's Law and Other Electrical Formulas

When a current flows in an electric circuit, the magnitude of the current is determined by dividing the electromotive force (volts, designated by the letter E) in the circuit by the resistance (ohms, designated by the letter R) of the circuit. The resistance is dependent on the material, cross section, and length of the conductor. The current is measured in amperes and is designated by the letter I. The relationship between an electric current (I), the electromotive force (E), and the resistance (R) is expressed by Ohm's law. The following equations take into consideration only pure resistance (i.e., not inductance or capacitance); therefore, they are customarily known as the DC formulas for Ohm's law. However, in most calculations for AC circuits, which is the ordinary wiring application, this formula is quite practical to use. Later in this book, another form of Ohm's law, dealing with inductive and capacitive reactance, will be discussed.

2-1   What are the three equations for Ohm's law and what do the letters in the formulas mean?

$$I = \frac{E}{R} \qquad R = \frac{E}{I} \qquad E = IR$$

Where: I is the current flow in amperes
E is the electromotive force in volts
R is the resistance in ohms

2-2   A direct-current circuit has a resistance of 5 ohms. If a voltmeter connected across the terminals of the circuit reads 10 volts, how much current is flowing?
From Ohm's law, the current is:

$$I = \frac{E}{R} = \frac{10}{5} = 2 \text{ amp}$$

2-3   If the resistance of a circuit is 25 ohms, what voltage is necessary for a current flow of 4 amperes?

From Ohm's law:

$$E = I \times R = 4 \times 25 = 100 \text{ volts}$$

2-4   If the potential across a circuit is 40 volts and the current is 5 amperes, what is the resistance?

From Ohm's law:

$$R = \frac{E}{I} = \frac{40}{5} = 8 \text{ ohms}$$

Series Circuits

A series circuit may be defined as one in which the resistive elements are connected in a continuous run (i.e., connected end to end) as shown in Fig. 2-1. It is evident that since the circuit has only one pathway (no branches), the amount of current flowing must be the same in all parts of the circuit. Therefore, the current flowing through each resistance is also equal. The total potential across the entire circuit equals the sum of potential drops across each individual resistance, or,

Fig. 2-1. Resistances in series.

$$E_1 = IR_1$$
$$E_2 = IR_2$$
$$E_3 = IR_3$$
$$E = E_1 + E_2 + E_3$$

and,

$$R = R_1 + R_2 + R_3$$

The equation for the total potential of the circuit is

$$E = IR_1 + IR_2 + IR_3 = I(R_1 + R_2 + R_3)$$

and,

$$I = \frac{E}{R_1 + R_2 + R_3} = \frac{E}{R}$$

2-5 If the individual resistances shown in Fig. 2-1 are 5, 10, and 15 ohms, respectively, what potential must the battery supply to force a current of 0.5 ampere through the circuit?

The total resistance is:

$$R = 5 + 10 + 15 = 30 \text{ ohms}$$

Hence, the total voltage is:

$$E = 0.5 \times 30 = 15 \text{ volts}$$

As a check, we can calculate the individual voltage drop across each part:

$$E_1 = 0.5 \times 5 = 2.5 \text{ volts}$$
$$E_2 = 0.5 \times 10 = 5.0 \text{ volts}$$
$$E_3 = 0.5 \times 15 = 7.5 \text{ volts}$$

and,

$$E = E_1 + E_2 + E_3 = 2.5 + 5.0 + 7.5 = 15 \text{ volts}$$

2-6 In order to determine the voltage of a DC source, three resistance units of 10, 15, and 30 ohms are conducted in series with this source. If the current through the circuit is 2 amperes, what is the potential of the source?

$$\begin{aligned} E &= I(R_1 + R_2 + R_3) \\ &= 2(10 + 15 + 30) = 2 \times 55 \\ &= 110 \text{ volts} \end{aligned}$$

Parallel Circuits

In a parallel, or divided, circuit such as that shown in Fig. 2-2, the same voltage appears across each resistance in the group; the current flowing through each resistance is inversely proportional to the value of the resistance. The sum of all the currents, however, is equal to the total current leaving the battery. Thus:

$$E = I_1 R_1 = I_2 R_2 = I_3 R_3$$

and,

$$I = I_1 + I_2 + I_3$$

$$I_T = I_1 + I_2 + I_3 \qquad R_T = \frac{1}{R_1} + \frac{1}{R_2} + \frac{1}{R_3} \qquad E_T = I_T + \frac{1}{R_T}$$

Fig. 2-2. Resistances in parallel.

When Ohm's law is applied to the individual resistance, the following equations are obtained:

$$I_1 = \frac{E}{R_1} \qquad I_2 = \frac{E}{R_2} \qquad I_3 = \frac{E}{R_3}$$

Hence,

$$I_1 = \frac{E}{R_1} + \frac{E}{R_2} + \frac{E}{R_3}$$

or

$$I = E\left(\frac{1}{R_1} + \frac{1}{R_2} + \frac{1}{R_3}\right)$$

and since $I = E/R$, the equivalent resistance of the several resistances connected in parallel is:

$$\frac{1}{R} = \frac{1}{R_1} + \frac{1}{R_2} + \frac{1}{R_3}$$

We have found, then, that any number of resistances in parallel can be replaced by an equivalent resistance whose value is equal to the reciprocal of the sum of the reciprocals of the individual resistances. You will find that the sum of resistances in parallel will always be smaller than the value of the smallest resistor in the group.

The value $1/R$, or the reciprocal of the value of the resistance, is expressed as the conductance of the circuit; its unit is mho, or ohm spelled backward, and is usually expressed by g or G.

Where there are only two resistances connected in parallel,

$$R = \frac{R_1 \times R_2}{R_1 + R_2} \text{ ohms}$$

Where there are any number of equal resistances connected in parallel, you may divide the value of one resistance by the number of equal resistances.

2-7 A resistance of 2 ohms is connected in series with a group of three resistances in parallel, which are 4, 5, and 20 ohms, respectively. What is the equivalent resistance of the circuit?

The equivalent resistance of the parallel network is:

$$\frac{1}{R} = \frac{1}{4} + \frac{1}{5} + \frac{1}{20} = .25 + .20 + .05 = .50$$

$$R = \frac{1}{.50} = 2 \text{ ohms}$$

The circuit is now reduced to two series resistors of 2 ohms each, as shown in Fig. 2-3; the equivalent resistance of the circuit is 2 + 2, or 4 ohms.

Fig. 2-3. Equivalent resistance of the series-parallel circuit of question 2-7.

2-8 Two parallel resistors of 2 and 6 ohms are connected in series with a group of three parallel resistors of 1, 3, and 6 ohms, respectively. If the two parallel-resistance groups are connected in series by means of a 1.5-ohm resistor, what is the equivalent resistance of the system?

Replace the 2- and 6-ohm resistors by a resistance of R_1, where:

$$R_1 = \frac{2 \times 6}{2 + 6} = \frac{12}{8} = 1.5 \text{ ohms}$$

Replace the group of three resistors by R_2, where:

$$\frac{1}{R} = \frac{1}{1} + \frac{1}{3} + \frac{1}{6} = 1.0 + 0.33 + 0.17 = 1.50$$

$$R = \frac{1}{1.50} = 1.67 \text{ ohms}$$

The circuit is now reduced to three series resistances, the values of which are 1.5, 1.5 and 0.67 ohms, as shown in Fig. 2-4. Their combined values are:

$$R_T = R + R_1 + R_2$$
$$R_T = 1.5 + 1.5 + 0.67 = 3.67 \text{ ohms}$$

Fig. 2-4. Equivalent resistance of the series-parallel circuit of question 2-8.

Units of Area and Resistance

The circular mil is the unit of cross section used in the American wire gauge (AWG) or the B&S wire gauge (see *NEC*, Table 8). The term *mil* means one thousandth of an inch (0.001 inch). A circular mil is the area of a circular wire with a diameter of 1 mil (0.001 inch).

The circular mil area of any solid cylindrical wire is equal to its diameter (expressed in mils) squared. For example, the area in circular mils (written CM or cir. mils) of a wire having a diameter of ⅜ inch (0.375) equals 375×375 = 375² = 140,625 CM. The diameter in mils of a solid circular wire is equal to the square root of its circular mil area. Assuming that a conductor has an area of 500,000 CM, its diameter in mils is the square root of 500,000, which is equal to 707 mils, or 0.707 inch (approximately). The area in square inches of a wire whose diameter is 1 mil is:

$$\frac{\pi}{4} = 0.7854 \times 0.001^2 = 0.0000007854 \text{ sq. in.}$$

The square mil is the area of a square whose sides are each 1 mil (0.001 inch). Hence, the area of a square mil is 0.001^2, or 0.000001 square inch. With reference to the previous definitions of the circular mil and the square mil, it is obvious that in order to convert a unit of circular area into its equivalent area in square mils, the circular mil must be multiplied by $\pi/4$, or 0.7854, which is the same as dividing by 1.273. Conversely, to convert a unit area of square mils into its equivalent area in circular mils, the square mil should be divided by $\pi/4$, or 0.7854, which is the same as multiplying by 1.273.

The above relations may be written as follows:

$$\text{Square mils} = \text{circular mils} \times 0.7854 = \frac{\text{circular mils}}{1.273}$$

$$\text{Circular mils} = \frac{\text{square mils}}{0.7854} = \text{square mils} \times 1.273$$

Thus, any circular conductor may be easily converted into a rectangular conductor (a bus bar, for example) containing an equivalent area of current-carrying capacity.

Fig. 2-5. Enlarged view of one circular mil and one square mil, with a comparison of the two.

2-9 If a No. 10 wire (B&S or AWG) has a diameter of 101.9 mils, what is its circular mil area?

$$\text{Area} = 101.9^2 = 10,383.6 \text{ CM}$$

Referring to Table 2-1, you will find this to be approximately true. If you concentrate on remembering the CM area of No. 10 wire as 10,380 CM, you will find this invaluable in arriving at the CM area of any size wire without the use of a table. For practical purposes, if you do not have a wire table such as Table 2-1 readily available, you may find the circular mil area of any wire size by juggling back and forth as follows, and the answer will be close enough for all practical purposes.

Every three sizes removed from No. 10 doubles or halves in CM
Every ten sizes removed from No. 10 is 1/10 or 10 times in CM

Example:
Ten sizes smaller than No. 10 is No. 0 wire. No. 10 wire has a circular mil area of 10,380 CM; therefore, No. 0 would be 103,800 CM. Table 2-1 shows an area of 105,600 CM for No. 0 wire, or about a 1.7% error.

Example:
Ten sizes smaller than No. 10 is No. 20 wire. Number 20 wire should then have a circular mil area of 1,038.1 CM.

You can see from the examples above that for quick figuring, the percentage of error is quite small; however, you should always have a wire table nearby. (You can find a pretty good one in chapter nine, table eight of *NEC*.)

2-10 Number 000 wire (AWG) has an area of 167,800 CM. If it were solid copper (not made of strands), what would its diameter be?

The diameter is:

$$\sqrt{167,800} = 409.6 \text{ mils}, \quad \text{or } 0.4096 \text{ inch}$$

2-11 A certain switchboard arrangement necessitates a conversion from a circular conductor to a rectangular bus bar having an equivalent area. If the diameter of the solid conductor measures 0.846 inch, calculate (a) the width of an equivalent bus bar, if the thickness of the bus bar is ¼ inch; (b) its area in circular mils; (c) the current-carrying capacity, if 1 square inch of copper carries 1,000 amperes.

The area, in square inches, of the circular conductor is:

$$A = 0.7854 \times D^2 = 0.7854 \times 0.846^2 = 0.562 \text{ square inch}$$

The area, in square inches, of the bus bar is:

$$A' = 0.25 \times W$$

where W is the width of the bus bar.
Since,

$$A' = A = 0.562 = 0.25 \, W$$

(a) $W = 0.562/025 = 2.248$ inches
(b) The area of the conductor $= 846^2 = 715,716$ CM
(c) The current-carrying capacity of the conductor =
$0.562 \times 1,000 = 562$ amps.

2-12 A certain 115-volt, 100-horsepower DC motor has an efficiency of 90% and requires a starting current of 150% times the full-load current. Determine (a) the size of fuses; (b) the copper requirements of the switch. Assume that 1 square inch of copper carries 1,000 amperes.

Motor current:

$$I_M = \frac{hp \times 746}{E \times efficiency} = \frac{100 \times 746}{115 \times 0.9} = 721 \text{ amp}$$

(a) The amperage of the fuses is, therefore, $721 \times 1.5 = 1,081.5$, or $1,200.0$ amperes, which is the rating of the closest manufactured fuse.

(b) Since the motor required a current of 721 amperes, the copper of each switch blade must be 721/1,000, or 0.721 square inch. Therefore, if ⅜-inch bus copper is used, its width must be:

$$W = \frac{0.721}{0.375} = 1.92 \text{ inches}$$

The mil-foot is a unit circular conductor that is one foot in length and one mil in diameter. The resistance of such a unit of copper has been found experimentally to be 10.37 ohms at 20°C; this is normally thought of as 10.4 ohms (Fig. 2-6).

A mil-foot of copper at 20°C offers 10.4 ohms resistance; at 30°C, it is 11.2 ohms; at 40°C, it is 11.6 ohms; at 50°C, it is 11.8 ohms; at 60°C, it is 12.3 ohms; and at 70°C, it is 12.7 ohms. Thus, in voltage-drop calculations, 12 is generally used as the constant K, unless otherwise specified, to allow for higher temperatures and to afford some factor of safety.

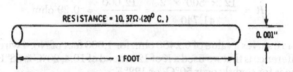

RESISTANCE = 10.37Ω (20° C.)

0.001"

1 FOOT

Fig. 2-6. Dimensions and resistance of one circular mil-foot of copper.

The resistance of a wire is directly proportional to its length and inversely proportional to its cross-sectional area. Therefore, if the resistance given in ohms of a mil-foot of wire is multiplied by the total length in feet (remember that there are practically always two wires, so if the distance is given in feet, multiply it by two to get the total resistance of both wires; this is a common error when working examinations) and divided by its cross-sectional area in circular mils, the result will be the total resistance of the wire in ohms. This is expressed as:

$$R = \frac{K \times L \times 2}{A}$$

Where: R is the resistance in ohms
  K is the constant (12) for copper
  L is the length in feet one way
  A is the area in circular mils

K for copper was given, so it will be well to give K for commercial aluminum:

20°C (68°F): K = 17.39 50°C (122°F): K = 19.73
30°C (86°F): K = 18.73 60°C (140°F): K = 20.56
40°C (104°F): K = 19.40 70°C (158°F): K = 21.23

K = 12 was used for copper, as in most instances it gives a safety factor in figuring voltage drop. Thus, with a correction factor of 1.672 for aluminum, as opposed to copper, it would be well to use a K factor of 20 for aluminum.

Should you have other conditions, K factors have been given for both copper and aluminum at various degrees Celsius. To change Celsius to Fahrenheit, use the following formula:

$$\text{Degrees C} \times 1.8 + 32 = \text{Degrees F}$$

Thus:

$$30°C \times 1.8 + 32 = 54 + 32$$
$$= 86°F$$

2-13 What is the resistance of a 500-foot line of No. 4 copper wire?

From Table 8 in the *NEC*, No. 4 wire has a cross-sectional area of 41,740 CM. Therefore,

$$R = \frac{12 \times 500 \times 2}{41,740} = \frac{12,000}{41,700} = 0.29 \text{ ohm}$$

Table 8 serves a value of 0.2480 ohm for 1,000 feet of No. 4 wire at 20°C. The difference is that we used a K of 12 instead of 10.4; 20°C is 68°F, and the 12 value is for slightly over 50°C, or 122°F.

2-14 Suppose it is desired to have a copper wire of 0.5-ohm resistance whose total length is 2,000 feet, or 1,000 feet one way. What must be its cross-sectional area? What size wire is necessary?

$$A = K\frac{L}{R} = \frac{12 \times 2,000}{0.5} = 48,000 \text{ CM}$$

According to Table 8, No. 4 wire has a cross-sectional area of 41,740 CM, and No. 3 wire has an area of 52,620 CM, so we would use the larger size No. 3 wire. Also, notice that the equation did not use the factor 2 because the 2,000-foot length was the total and not just the length of one wire.

2-15 If the resistance of a copper wire whose diameter is ⅛ inch is measured as 0.125 ohm, what is the length of the wire?

$$L = \frac{RA}{K} = \frac{0.125 \times 125^2}{12} = 163 \text{ ft}$$

Here again, this is the total length; one way would be 81.5 feet.

2-16 A copper line that is 5 miles in length has a diameter of 0.25 inch. Calculate: (a) the diameter of the wire in mils; (b) the area of the wire in circular mils; (c) the weight in pounds; (d) the resistance at 50°C. Assume that the weight in pounds per cubic inch is 0.321.

    (a)   The diameter in mils = $1,000 \times 0.25 = 250$ mils.

    (b)   The area in CM = $250^2 = 62,500$ CM

    (c)   Cross-sectional area = $0.7854 \times D^2 = 0.7854 \times 0.25^2 = 0.0491$ sq. in.
Length of wire = $5,280 \times 5 \times 2 \times 12 = 633,600$ in.
Weight of wire = $0.0491 \times 633,600 \times 0.321 = 9,986.23$, or about 10,000 lb.

    (d) $$R = \frac{K \times 2 \times (5 \times 5,280)}{62,500}$$

$$= \frac{12 \times 2 \times 26,400}{62,500}$$

$$= 10.14 \text{ ohms}$$

Skin Effect

When alternating current flows through a conductor, an inductive effect occurs, which tends to force the current to the surface of the conductor. This produces a voltage loss and also affects the current-carrying capacity of the conductor. For open wires or wires in nonmetallic-sheathed cable, this effect is neglected until the No. 0 wire size is reached. In metallic-sheathed cables and metallic raceways, the skin effect is neglected until size No. 2 is reached. At these points, there are multiplying factors for conversion from DC resistance to AC resistance (Table 2-1). Note that there is a different factor for aluminum than for copper cables.

2-17 The DC resistance of a length of 250,000-CM copper cable in conduit (rigid metal) was found to be 0.05 ohm. What would its AC resistance be?

From Table 2-1, the multiplying factor is found to be 1.06, therefore, $R_{AC} = 0.05$ ohm $\times 1.06 = 0.053$ ohm.

Table 2-1. Multiplying Factors for Converting DC Resistance to 60-Cycle AC Resistance

Size	Multiplying Factor			
	For Nonmetallic-Sheathed Cables in Air or Nonmetallic Conduit		For Metallic-Sheathed Cables or All Cables in Metallic Raceways	
	Copper	Aluminum	Copper	Aluminum
Up to 3 AWG	1.000	1.000	1.00	1.00
2	1.000	1.000	1.01	1.00
1	1.000	1.000	1.01	1.00
0	1.001	1.000	1.02	1.00
00	1.001	1.001	1.03	1.00
000	1.002	1.001	1.04	1.01
0000	1.004	1.002	1.05	1.01
250 MCM	1.005	1.002	1.06	1.02
300 MCM	1.006	1.003	1.07	1.02
350 MCM	1.009	1.004	1.08	1.03
400 MCM	1.011	1.005	1.10	1.04
500 MCM	1.018	1.007	1.13	1.06
600 MCM	1.025	1.010	1.16	1.08
700 MCM	1.034	1.013	1.19	1.11
750 MCM	1.039	1.015	1.21	1.12
800 MCM	1.044	1.017	1.22	1.14
1000 MCM	1.067	1.026	1.30	1.19
1250 MCM	1.102	1.040	1.41	1.27
1500 MCM	1.142	1.058	1.53	1.36
1750 MCM	1.185	1.079	1.67	1.46
2000 MCM	1.233	1.100	1.82	1.56

Voltage-Drop Calculations

We have discussed the methods for finding the resistance of wire. By the use of Ohm's law, we can now find the voltage drop for circuits loads. Use the form,

$$E = I \times R$$

Under Section 215-2(b), FPN 2 of the *NEC*, we find the prescribed maximum allowable percent of voltage drop permitted for feeders and branch circuits. In Section 210-19(a), FPN 4 of the *NEC*, the maximum allowable percent of voltage drop for branch circuits is given.

2-18 What is the percentage of allowable voltage drop for feeders that are used for power and heating loads?
Maximum of 3%.

2-19 What is the percentage of allowable voltage drop for feeders that are used for lighting loads?
Maximum of 1 to 3%.

2-20 What is the percentage of allowable voltage drop for combined lighting, heating, and power loads?
Maximum of 1 to 3% for feeders and 5% for feeders and branch circuits. The following formula is used for voltage drop calculations.

Since,

$$R = \frac{K \times L \times 2}{A}$$

then,

$$E_d = I \times R = \frac{K \times 2L \times I}{A}$$

Where:

E_d is the voltage drop of the circuit
$2L$ is the total length of the wire
K is a constant (12)
I is the current, in amperes, of the circuit
A is the area, in circular mils, of the wire in the circuit

By transposing the formula above, we can determine the circular mil area of a wire for a specified voltage drop:

$$A = \frac{K \times 2L \times I}{E_d}$$

2-21 A certain motor draws 22 amps at 230 volts, and the feeder circuit is 150 feet in length. If No. 10 copper wire is desired, what would be the voltage drop? Would No. 10 wire be permissible to use?

$$E_d = \frac{12 \times 2 \times 150 \times 22}{10,380}$$

$$= \frac{79,200}{10,380}$$

$$= 7.63 \text{ volts}$$

However, $230 \times 0.03 = 6.90$ volts, which is the voltage drop permissible under Section 215-2 of the *NEC*, so No. 10 wire would not be large enough.

2-22 In question 2-21, a voltage drop of 7.63 volts was calculated; however, the maximum permissible voltage drop is 6.9 volts. What size wire would have to be used?

According to Table 8 in the *NEC*, No. 10 wire has an area of 10,380 CM, and No. 9 wire has an area of 13,090 CM. Therefore, No. 9 would be the proper size, except that you cannot purchase No. 9 wire; you will have to use No. 8 wire, which has an area of 16,510 CM.

2-23 If No. 8 copper wire were used in question 2-22, what would the voltage drop be? What percentage would this drop be?

$$E_d = \frac{12 \times 2 \times 150 \times 22}{16,509} = \frac{79,200}{16,509} = 4.80 \text{ volts}$$

$$\frac{4.80}{230} \times 100 = 2.09\%$$

$$A = \frac{K \times 2L \times I}{E_d}$$

$$= \frac{12 \times 2 \times 150 \times 22}{6.9}$$

$$= 11,478.3 \text{ CM}$$

Formulas for Determining Alternating Current in Alternating-Current Circuits

In the formulas of Fig. 2-7:

R is the resistance in ohms
X_L is the inductive reactance in ohms = $2\pi f L$
X_C is the capacitive reactance in ohms = $2\pi f C$
f is the frequency
L is the inductance in henrys
C is the capacity in farads
Z is the impedance in ohms
I is the current in amperes
E is the pressure in volts

Fig. 2-7. Fundamental forms of AC circuits, with the method of determining current when voltage and impedance are known.

Formulas for Combining Resistance and Reactance

In the following formulas of Fig. 2-8:

R is the resistance in ohms
X_L is the inductive reactance in ohms = $2\pi fL$
X_C is the capacitive reactance in ohms = $1/2\pi fC$
f is the frequency
L is the inductance in henrys
C is the capacity in farads
Z is the impedance in ohms
I is the current in amperes
E is the pressure in volts

Fig. 2-8. Methods of impedance determination in AC circuits when resistance, inductive reactance, and capacitive reactance (in ohms) are known.

2-24 A coil has a resistance of 10 ohms and an inductance of 0.1 henry. If the frequency of the source is 60 hertz, what is the voltage necessary to cause a current of 2 amperes to flow through the coil?

With reference to the previously derived formula,

$$E_R = IR = 2 \times 10 = 20 \text{ volts}$$
$$X_L = 2\pi fL = 2\pi \times 60 \times 0.1 = 37.7 \text{ ohms}$$
$$E_L = IX_L = 2 \times 37.7 = 75.4 \text{ volts}$$

The applied voltage must therefore be:

$$E = \sqrt{E_R{}^2 + E_L{}^2} = \sqrt{20^2 + 75.4^2} = 78 \text{ volts}$$

2-25 A coil with a negligible resistance requires 3 amperes when it is connected to a 180-volt, 60-hertz supply. What is the inductance of the coil?

$$X_L = \frac{E}{I} = \frac{180}{3} = 60 \text{ ohms}$$

and,

$$X_L = 2\pi fL$$

therefore,

$$L = \frac{60}{2\pi \times 60} = 0.159 \text{ henry}$$

2-26 An alternating current of 15 amperes with a frequency of 60 hertz is supplied to a circuit containing a resistance of 5 ohms and an inductance of 15 millihenrys. What is the applied voltage?

$$X_L = 2\pi fL = 2\pi \times 60 \times 0.015 = 5.65 \text{ ohms}$$
$$Z = \sqrt{R^2 + X_L{}^2} = \sqrt{5^2 + 5.65^2} = 7.54 \text{ ohms}$$
$$E = IZ = 15 \times 7.54 = 113.1 \text{ volts}$$

2-27 A coil contains 5 ohms resistance and 0.04 henry inductance. The voltage and frequency of the source are 100 volts at 60 cycles. Find (a) the impedance of the coil; (b) the current through the coil; (c) the voltage drop across the inductance; (d) the voltage drop across the resistance

$$X_L = 2\pi fL = 2\pi \times 60 \times 0.04 = 15 \text{ ohms}$$

(a) $$Z = \sqrt{5^2 + 15^2} = \sqrt{250} = 15.8 \text{ ohms}$$

(b) $$I = \frac{E}{Z} = \frac{100}{\sqrt{5^2 + 15^2}} = 6.3 \text{ amperes}$$

(c) $E_L = IX_L = 6.3 \times 15 = 94.5$ volts

(d) $E_R = IR = 6.3 \times 5 = 31.5$ volts

2-28 An alternating-current contains 10 ohms resistance in series with a capacitance of 40 microfarads. The voltage and frequency of the source are 120 volts at 60 cycles. Find (a) the current in the circuit; (b) the voltage drop across the resistance; (c) the voltage drop across the capacitance; (d) the power factor; (e) the power loss.

$$X_C = \frac{1}{2\pi fC} = \frac{1}{2\pi \times 60 \times 0.00004} = 66.67 \text{ ohms}$$

$$Z = \sqrt{10^2 + 66.7^2} = 67 \text{ ohms}$$

(a) $I = \dfrac{E}{Z} = \dfrac{120}{67} = 1.8$ amperes

(b) $E_R = IR = 1.8 \times 10 = 18$ volts

(c) $E_C = IX_C = 1.8 \times 66.3 \times 119.3$ volts

(d) $\cos \phi = \dfrac{R}{Z} = \dfrac{10}{67} = 0.15$, or 15%

(e) $P = E \times I \times \cos \phi$

$\qquad = 120 \times 1.8 \times 0.15$

$\qquad = 32.4$ watts

2-29 A coil of 3 ohms resistance and 20 millihenrys inductance is connected in series with a capacitance of 400 microfarads. If the voltage and frequency are 120 volts at 60 hertz, find (a) the impedance of the circuit; (b) the current in the circuit; (c) the power loss; (d) the power factor.

$$X_L = 2\pi fL = 2\pi \times 60 \times 0.020 = 7.54 \text{ ohms}$$

$$X_C = \frac{1}{2\pi \times 60 \times 0.0004} = 6.67 \text{ ohms}$$

(a) $Z = \sqrt{R^2 + (X_L - X_C)^2}$

$\qquad = \sqrt{3^2 + (7.54 - 6.67)^2}$

$\qquad = 3.12$ ohms

(b)
$$I = \frac{E}{Z} = \frac{120}{3.12} = 38.5 \text{ amperes}$$

(c)
$$P = I^2R = 38.5^2 \times 3 = 4{,}447 \text{ watts}$$

(d)
$$\cos \phi = \frac{R}{Z} = \frac{3}{3.12} = 0.961, \text{ or } 96\% \text{ (approx.)}$$

As a check for the power loss, in part (c), use the information obtained in part (d), the power factor. Then:

$$P = EI \cos \phi = 120 \times 38.5 \times 0.961 = 4{,}440 \text{ watts}$$

2-30 A certain series circuit has a resistance of 10 ohms, a capacitance of 0.0003 farad, and an inductance of 0.03 henry. If a 60-hertz, 230-volt emf is applied to this circuit, find (a) the current through the circuit; (b) the power factor; (c) the power consumption.

$$X_L = 2\pi f L = 2\pi \times 60 \times 0.03 = 11.30 \text{ ohms}$$

$$X_C = \frac{1}{2\pi f C} = \frac{1}{2\pi \times 60 \times 0.0003} = 8.85 \text{ ohms}$$

$$Z = \sqrt{R^2 + (X_L - X_C)^2}$$
$$= \sqrt{10^2 + (11.30 - 8.85)^2}$$
$$= 10.3 \text{ ohms}$$

(a)
$$I = \frac{E}{Z} = \frac{230}{10.3} = 22.3 \text{ amperes}$$

(b)
$$\cos \phi = \frac{10}{10.2} = 0.97, \text{ or } 97\%$$

(c)
$$P = I^2R = 22.3^2 \times 10 = 4.97 \text{ kilowatts}$$

2-31 The circuit in Fig. 2-9 contains a resistance of 30 ohms and a capacitance of 125 microfarads. If an alternating current of 8 amperes at a frequency of 60 hertz is flowing in the circuit, find (a) the voltage drop across the resistance; (b) the voltage drop across the capacitance; (c) the voltage applied across the circuit.

$$X_C = \frac{1}{2\pi f C} = \frac{1}{2\pi \times 60 \times 0.000125} = 21.3 \text{ ohms}$$

$$Z = \sqrt{R^2 + X_C^2} = \sqrt{30^2 + 21.3^2} = 36.8 \text{ ohms}$$

(a)
$$E_R = IR = 8 \times 30 = 240 \text{ volts}$$

(b)
$$E_C = IX_C = 8 \times 21.3 = 170 \text{ volts}$$

(c)
$$E = IZ = 8 \times 36.8 = 294 \text{ volts}$$

Fig. 2-9. Resistance and capacitance in series, with appropriate vector diagrams.

2-32 A resistance of 15 ohms is connected in series with a capacitance of 50 microfarads. If the voltage of the source is 120 volts at 60 hertz, find (a) the amount of current in the circuit; (b) the voltage drop across the resistance; (c) the voltage drop across the capacitance; (d) the angular difference between the current and the applied voltage; (e) the power loss in the circuit.

$$X_C = \frac{1}{2\pi f C} = \frac{1}{2\pi \times 60 \times 0.00005} = 53.0 \text{ ohms}$$

$$Z = \sqrt{R^2 + X_C^2} = \sqrt{15^2 + 53.0^2} = 55.1 \text{ ohms}$$

(a) $\quad I = \dfrac{E}{Z} = \dfrac{120}{55.1} = 2.18 \text{ amperes}$

(b) $\quad E_R = IR = 2.18 \times 15 = 32.7 \text{ volts}$

(c) $\quad E_C = IX_C = 2.18 \times 53.0 = 115.5 \text{ volts}$

(d) $\quad \cos \phi = \dfrac{E_R}{E} = \dfrac{32.7}{120} = 0.272, \text{ and } \phi = 74.2°$

(e) $\quad P = I^2 R = 2.18^2 \times 15 = 71 \text{ watts}$

2-33 A resistance of 20 ohms and a capacitance of 100 microfarads are connected in series across a 200-volt, 50-hertz AC supply. Find (a) the current in the circuit; (b) the potential drop across the resistance; (c) the potential drop across the capacitance; (d) the phase difference between the current and the applied voltage; (e) the power consumed; (f) the power factor.

$$X_C = \frac{1}{2\pi f C} = \frac{1}{2\pi \times 50 \times 0.0001} = 31.8 \text{ ohms}$$

$$Z = \sqrt{R^2 + X_C^2} = \sqrt{20^2 + 31.8^2} = 37.6 \text{ ohms}$$

(a) $\qquad I = \dfrac{E}{Z} = \dfrac{200}{37.6} = 5.32$ amperes

(b) $\qquad E = IR = 5.32 \times 20 = 106.4$ volts

(c) $\qquad Ec = IX_C = 5.32 \times 31.8 = 169.2$ volts

(d) $\qquad \cos \phi = \dfrac{R}{Z} = \dfrac{20}{37.6} = 0.532, \quad$ and $\phi = 57.9°$

(e) $\qquad P = I^2R = 5.32^2 \times 20 = 566$ watts

(f) $\qquad \cos \phi = 0.532, \quad$ or 53.2%

2-34 What is the total capacitance of four parallel capacitors that are 10, 15, 25 and 30 microfarads, respectively?

$$\begin{aligned} C &= C_1 + C_2 + C_3 + C_4 \\ &= 10 + 15 + 25 + 30 \\ &= 80 \text{ microfarads} \end{aligned}$$

Fig. 2-10. Resistance, inductance, and capacitance in series, with a vector diagram illustrating their relationship to each other.

2-35 A certain coil (shown in Fig. 2-10) with a resistance of 5 ohms and an inductance of 0.01 henry is connected in series with a capacitor across a 10-volt supply, which has a frequency of 800 cycles per second. Find (a) the capacitance that will produce resonance; (b) the corresponding value of the current; (c) the potential drop across the coil; (d) the potential drop across the capacitor; (e) the power factor of the circuit; (f) the power consumption.

The inductive reactance of the coil is

$$X_L = 2\pi \times 800 \times 0.01 = 50.24 \text{ ohms}$$

Since resonance occurs when $X_L = X_C$, X_C must also be equal to 50.24 ohms. Therefore,

(a) $\qquad X_C = 50.24 = \dfrac{10^6}{2\pi \times 800 = C}$

$\qquad\qquad C = \dfrac{10^4}{50.24^2}$

$\qquad\qquad\quad = 3.96 \text{ microfarads}$

(b) At resonance, the current is:

$$I = \frac{E}{R} = \frac{10}{5} = 2 \text{ amperes}$$

(c) The potential drop across the coil is:

$$E_L = IX_L = 2 \times 50.24 = 100.5 \text{ volts}$$

(d) The potential drop across the capacitor is:

$$E_C = IX_C = 2 \times 50.24 = 100.5 \text{ volts}$$

(e) The power factor is:

$$\cos \phi = \frac{R}{Z}$$

but since resonance $Z = R$,

$$\cos \phi = \frac{R}{R} = 1, \quad \text{and } \phi = 0°$$

(f) The power consumed is:

$$P = I^2R = 4 \times 5 = 20 \text{ watts}$$

2-36 **The field winding of a shunt motor has a resistance of 110 ohms, and the emf applied to it is 220 volts. What is the amount of power expended in the field excitation?**

The current through the field is:

$$I_f = \frac{E_t}{R_f} = \frac{220}{110} = 2 \text{ amperes}$$

The power expended = $E_tI_f = 220 \times 2 = 440$ watts. The same results can also be obtained directly by using the following equation:

$$P_f = \frac{E_f^2}{R_f} = \frac{220^2}{110} = 440 \text{ watts}$$

2-37 **A shunt motor whose armature resistance is 0.2 ohm and whose terminal voltage is 220 volts requires an armature current of 50 amperes and runs at 1,500 rpm when the field is fully excited. If the strength of the field is decreased and the amount of armature current is increased, both by 50%, at what speed will the motor run?**

The expression for the counter emf of the motor is:

$$E_a = E_t - I_aR_a$$

and,

$$E_{a1} = 220 - (50 \times 0.2) = 210 \text{ volts}$$

similarly,

$$E_{a2} = 220 - (75 \times 0.2) = 205 \text{ volts}$$

also, $\qquad E_a = N \uparrow K$

and, $\qquad \dfrac{E_{a1}}{E_{a2}} = \dfrac{N_1 \phi_1 K_1}{N_2 \phi_2 K_2}$

Since the field is decreased by 50%,

$$\phi_1 = 1.5\phi_2, \quad \text{and } Z_1 = Z_2$$

it follows that:

$$\frac{210}{205} = \frac{1,500 \times 1.5}{N_2}$$

$$N_2 = \frac{1,500 \times 205 \times 1.5}{210} = 2,196 \text{ rpm}$$

2-38 **A 7.5-hp, 220-volt interpole motor has armature and shunt-field resistances of 0.5 ohm and 200 ohms, respectively. The current input at 1,800 rpm under no-load conditions is 3.5 amperes. What are the current and electromagnetic torque for a speed of 1,700 rpm?**
Under no-load conditions (at 1,800 rpm),

$$I_a = I_L - I_f = 3.5 - \frac{220}{200} = 2.4 \text{ amp}$$

$$\phi K_{NL} = \frac{E_t - (I_a R_a)}{N} = \frac{220 - (2.4 \times 0.5)}{1,800} = 0.1216$$

$$\phi K_{NL} = \phi K_{FL}$$

At 1,700 rpm,

$$
\begin{aligned}
I_a &= \frac{\dot{E}_t - (N\phi K)}{R_a} \\
&= \frac{220 - (1,700 \times 0.1216)}{0.5} \\
&= 26.6 \text{ amp}
\end{aligned}
$$

$$I_L = I_a + I_f = 26.6 + 1.1 = 27.7 \text{ amp}$$

$$T_e = 7.05\phi K I_a = 7.05 \times 0.1216 \times 26.6 = 22.8 \text{ ft-lb}$$

2-39 **The mechanical efficiency of a shunt motor whose armature and field resistances are 0.055 and 32 ohms, respectively, is to be tested by means of a rope brake. When turning at 1,400 rpm, the longitudi-**

nal pull on the 6-inch-diameter pulley is 57 lb. Simultaneous readings on the line voltmeter and ammeter are 105 volts and 35 amperes, respectively. Calculate (a) the counter emf developed; (b) the copper losses; (c) the efficiency.

$$I_a = I_L - I_f = 35 - \frac{105}{32} = 31.7 \text{ amp}$$

(a) $$E_a = E_t - (I_aR_a) = 105 - (31.7 \times 0.055)$$
$$= 103.26 \text{ volts}$$

(b) $$P_C = I_f{}^2R_f + I_a{}^2R_a$$
$$= (3.3^2 \times 32) + (31.7^2 \times 0.055)$$
$$= 404 \text{ watts}$$

(c) $$\text{Output} = \frac{2p \times 1,400 \times {}^3/_{12} \times 57}{33,000} = 3.8 \text{ hp}$$

$$\text{Input} = \frac{105 \times 35}{746} = 4.93 \text{ hp}$$

$$\eta_m = \frac{3.8}{4.93} = 0.771, \quad \text{or } 77\%$$

2-40 A copper transmission line that is 1.5 miles in length is used to transmit 10 kilowatts from a 600-volt generating station. The voltage drop in the line is not to exceed 10% of the generating-station voltage. Calculate (a) the line current; (b) the resistance of the line; (c) the cross-sectional area of the wired.

(a) $$I_L = \frac{10,000}{600} = 16.67 \text{ amperes}$$

The permissible voltage drop = $600 \times 0.1 = 60$ volts

(b) $$R = \frac{60}{16.67} = 3.6 \text{ ohms}$$

(c) $$3.6 = \frac{10.4 \times 3 \times 5,280}{A}$$

$$A = \frac{10.4 \times 3 \times 5,280}{3.6} = 45,760 \text{ CM}$$

2-41 A trolley system ten miles long is fed by two substations that generate 600 volts and 560 volts, respectively. The resistance of the trolley wire and rail return is 0.3 ohm per mile. If a car located four miles from the 600-volt substation draws 200 amperes from the line, what is the voltage between the trolley collector and track? How much current is supplied by each substation?

With reference to Fig. 2-11, the following equation can be written:

Equation (1) $I_1 + I_2 = 200$ amperes

That is, the arithmetical sum of the current drain from each substation must equal the current drawn by the trolley car. In a similar manner, the equations for the voltage drop in each branch of the trolley wire are:

Equation (2) $I_1(1.2) = 600 - E$
Equation (3) $I_2(1.8) = 560 - E$

Subtracting equation (3) from equation (2),

Equation (4) $40 = 1.2I_1 - 1.8I_2$

According to equation (1),

$$I_1 = 200 - I_2$$

Therefore, equation (4) becomes

$$40 = 1.2(200 - I_2) - 1.8I_2$$
$$I_2 = 66.67 \text{ amperes}$$
$$I_1 = 200 - 66.67 = 133.33 \text{ amperes}$$

Fig. 2-11. Current and potential drop in a trolley-feeder system.

By inserting the value of I_2 in equation (3), we obtain the voltage between the trolley collector and the track:

$$E = 560 - (1.8 \times 66.67) = 440 \text{ volts}$$

The same result can be obtained by inserting the value of I_1 in equation (2).

2-42 It is desired to supply power from a 220-volt source to points C and D in Fig. 2-12 by means of the feeder arrangement indicated. The motor at point C requires 120 amperes, and the motor at point D requires 80 amperes. With the length of the wires as indicated and a maximum voltage drop of 10%, calculate (a) the cross-sectional area of feeder AB; (b) the cross-sectional area of feeder BC; (c) the cross-sectional area of feeder BD; (d) the power loss in each section.

Fig. 2-12. Branch feeder calculations.

The voltage drop across AC and AD is

$$E' = 220 \times 0.1 = 22 \text{ volts}$$

To simplify our calculation, the voltage drop across BC and BD can be arbitrarily set at 10 volts. The voltage drop across AB is, therefore, 22–10, or 12 volts.

(a) $\quad A = \left(\dfrac{10.4 \times 2}{12} \right) \times 200 \times 1,000 = 346,667 \text{ CM}$

(b) $\quad A = \left(\dfrac{10.4 \times 2}{10} \right) \times 120 \times 328 = 81,869 \text{ CM}$

(c) $\quad A = \left(\dfrac{10.4 \times 2}{10} \right) \times 80 \times 656 = 109,158 \text{ CM}$

(d) $\quad P_{AB} = 200 \times 12 = 2,400 \text{ watts}$
$\qquad P_{BC} = 120 \times 10 = 1,200 \text{ watts}$
$\qquad P_{BD} = 80 \times 10 = 800 \text{ watts}$

2-43 The motor illustrated in Fig. 2-13 is located at a distance of 500 feet from the generator and requires 40 amperes at 220 volts. No. 4 AWG wire is used. Calculate (a) the voltage at the generator; (b) the voltage-drop percentage in the line; (c) the power loss in the line; (d) the power-loss percentage; (e) the cost of power losses per year. Assume that the motor operates 8 hours per day, 300 days per year, at a cost of 3 cents per kilowatt-hour.

Fig. 2-13. Calculations for a 220-volt motor.

With reference to Table 2-1, the cross-sectional area of No. 4 wire = 41,740 CM.

$$R = \frac{10.4 \times 1,000}{41,740} = 0.25 \text{ ohms}$$

(a) $\quad E_G = 220 + (40 \times 0.25) = 230 \text{ volts}$

(b) $\quad \dfrac{(230 - 220)\,100}{220} = 4.55\%$

(c) $\quad P_G - P_R = 40^2 \times 0.25 = 400 \text{ watts}$

(d) $\quad \dfrac{(P_G - P_R)\,100}{P_R} = \dfrac{400 \times 100}{40 \times 220} = 4.55\%$

(e) $\quad$ Yearly cost of power losses $= 0.4 \times 8 \times 300 \times 0.03$
$$= \$28.80$$

2-44 Energy is transmitted from a switchboard to the combined load shown in Fig. 2-14. The lamp group requires 20 amperes, and the motor requires 30 amperes from the line. Number 2 wire (resistance = 0.156 ohm per 1,000 ft) is used throughout the circuit. Calculate (a) the power drawn by the lamps; (b) the power drawn by the motor; (c) the power loss in the line; (d) the total power supplied by the switchboard.

Fig. 2-14. Voltage-drop calculations in a feeder circuit which is supplying a motor and lamp load.

The resistance in line ABCD is:

$$\frac{0.162 \times 200 \times 2}{1,000} = 0.065 \text{ ohm}$$

Voltage across the lamps = $125 - (50 \times 0.065)$
$$= 121.8 \text{ volts}$$

(a) $\qquad P = 20 \times 121.8 = 2.436 \text{ kilowatts}$

The resistance in line BEFC is:

$$\frac{0.162 \times 100 \times 2}{1,000} = 0.0324 \text{ ohm}$$

Voltage at the motor = $121.8 - (30 \times 0.0324)$
$$= 120.83 \text{ volts}$$

(a) $\qquad P_M = 30 \times 120.83 = 3.62 \text{ kw}$

(b) $\qquad P_L = (50 \times 125) - (2,436 + 3,630) = 184 \text{ watts}$

(d) $\qquad P_T = 125 \times 50 = 6.25 \text{ kilowatts}$

Relative Conductivity

2-45 What is the comparison of electrical conductivity between silver, copper, and aluminum?
Silver is the best conductor of electricity of the three and is considered as possessing 100% conductivity. Copper is next and is considered as approximately 94% of the conductivity of silver. Aluminum is considered last and has approximately 61% of the conductivity of silver.

2-46 What are the advantages and disadvantages of silver as a conductor?
The price of silver makes it generally prohibitive to use for a conductor; it is only used in special cases for its high conductivity where the price is of no consequence.

2-47 What are the advantages of copper as a conductor? What are its disadvantages?
Copper is plentiful, relatively cheap, a good conductor, and it has a high tensile strength. It can be corroded under certain circumstances.

2-48 What are the advantages and disadvantages of aluminum as a conductor?
Aluminum is inexpensive, lightweight, and readily available. It corrodes more easily than copper, so its use is limited to some degree; it also has less tensile strength than copper.

CHAPTER 3

Power and Power Factor

The unit of *work* is the foot-pound. This is the amount of work done when a force of one pound acts through a distance of one foot. The amount of work done is equal to the force in pounds times the distance in feet, or:

$$W \text{ (work)} = \text{force} \times \text{distance}$$

Thus, if an object weighing 10 pounds is lifted a distance of 4 feet, the work done is equal to 40 ft-lb.

Time becomes involved when performing work, so we use the quantity known as *power*, which is the rate of doing work. Power is directly proportional to the amount of work done and is inversely proportional to the time in which the work is done. For example, more power is required if an object weighing 10 pounds is lifted through a distance of 4 feet in 1 minute than if the same 10-pound object is lifted a distance of 4 feet in 5 minutes. From this statement, we may arrive at formula for power:

$$\text{Power (foot - pounds per minute)} = \frac{\text{work done (foot - pounds)}}{\text{time (minutes)}}$$

The more common unit of power is the *horsepower*, which is equivalent to 33,000 ft-lb of work per minute. Remember this figure and its relationship to time and horsepower; it will be used quite often when working problems that deal with power.

In electricity, the unit of power is the watt. However, since the watt is a relatively small unit, the kilowatt is more commonly used for the unit of power. One kilowatt is equivalent to 1,000 watts.

A good working knowledge of the electrical formulas that are used to determine power is a *must* for the electrician. When using electrical formulas to determine power, it is a universal practice to use the following notations:

P is the power in watts
I is the current in amperes
R is the resistance in ohms
E is the potential difference in volts

Thus, the power P expended in a load resistance R when a current I flows due to a voltage pressure E can be found by the following relationships:

$$P = IE$$

$$P = \frac{E^2}{R}$$

$$P = I^2R$$

Remember, IR equals a potential, in volts, and I^2R equals power, in watts.

When dealing with large amounts of electrical power, it may be required that you be able to determine the cost of the power consumed. You will be dealing with units of kilowatts and kilowatt-hours (kwh), which means the number of kilowatts used per hour. Thus, 25 kwh is equivalent of 25 kilowatts used for 1 hour. To find the cost of an electrical-usage bill, the following formula is used:

$$\text{Cost} = \frac{\text{watts} \times \text{hours used} \times \text{rate per kwh}}{1,000}$$

For example, an electric heater that draws 1,350 watts is used for 4 hours, and the cost of electricity for that particular location is 3 cents per kilowatt-hour. What is the cost of using the heater?

$$\text{Cost} = \frac{1,350 \times 4 \times 0.03}{1,000} = 16.2 \text{ cents}$$

Energy can be changed from one form to another but can never be destroyed. Therefore, we may readily change electrical power into mechanical power; the converse is also true. The usual method of referring to mechanical power is in terms of units of horsepower; one horsepower is equal to 746 watts. This equality is valid when considering that the equipment used to produce one horsepower operates at 100% efficiency, which, of course, is not possible since there is always some power lost in the form of friction or other losses, which will be covered later in this text.

3-1   A motor draws 50 amperes and is fed by a line that is 125 feet long; the line consists of No. 6 copper wire. What is the I^2R loss of the line?

From Table 8 in the *NEC*, it is found that No. 6 copper has a resistance of 0.410 ohm per 1,000 feet at 25°C. The line is 125 feet long; therefore, the amount of wire used will be 2×125, or 250 feet. This is 25% of 1,000 feet, so the resistance of the wire will be 0.410 × 0.25, or 0.1025 ohm. Therefore,

$$P = I^2R = (50)^2 \times 0.1025 = 2,500 \times 0.1025 = 256.25 \text{ watts}$$

3-2 **The line loss of question 3-1 is 256.25 watts. If the motor is operated for 100 hours, with the rate of electricity being 3 cents per kwh, what would be the cost of the I^2R loss of the line?**

$$\text{Cost} = \frac{256.25 \times 100 \times 0.03}{1,000} = 76.875 \text{ cents}$$

3-3 **A 1-hp motor draws 1,000 watts. What is its efficiency?**

$$\text{Efficiency} = \frac{\text{output}}{\text{input}}$$

Therefore,

$$\text{Efficiency} = \frac{746}{1,000} = 0.746, \text{ or } 74.6\%$$

3-4 **An electric iron draws 11 amperes at 120 volts. How much power is used by the iron?**

$$P = 11 \times 120 = 1,320 \text{ watts} = 1.32 \text{ kw}$$

3-5 **A motor must lift an elevator car weighing 2,000 pounds to a height of 1,000 feet in 4 minutes. (a) What is the theoretical size, in horsepower, of the motor required? (b) At 50% efficiency, what is the size, in horsepower, of the motor required?**

(a)
$$W = 2,000 \times 1,000 = 2,000,000 \text{ ft - lb}$$

$$\frac{2,000,000}{4} = 500,000 \text{ ft - lb per minute}$$

$$\frac{500,000}{33,000} = 15.15 \text{ hp}$$

(b)
$$\text{Input} = \frac{\text{output}}{\text{efficiency}} = \frac{15.15}{0.50} = 30.3 \text{ hp}$$

A 30-hp motor will carry this load nicely.

3-6 **A lamp operating at 120 volts has a resistance of 240 ohms. What is the wattage of the lamp?**

$$P = \frac{E^2}{R} = \frac{120^2}{240} = \frac{14,400}{240} = 60 \text{ watts}$$

3-7   What is the overall efficiency of a 5-hp motor that draws 20 amperes at 240 volts?

$$\text{Input} = 240 \times 20 = 4{,}800 \text{ watts}$$

$$\text{Output} = 5 \times 746 = 3{,}730 \text{ watts}$$

$$\text{Efficiency} = \frac{3{,}730}{4{,}800} = 0.777, \text{ or } 77.7\%$$

3-8   What is the cost of operating a 2-watt electric clock for one year at 2 cents per kwh?

$$\text{Cost} = \frac{2 \text{ watts} \times 24 \text{ hours} \times 365 \text{ days} \times 0.02}{1{,}000} = 35.04 \text{ cents}$$

3-9   The primary of a transformer draws 4 amperes at 7,200 volts. A reading at the secondary shows 110 amperes at 240 volts. What is the efficiency of the transformer at this load?

$$\text{Efficiency} = \frac{\text{output}}{\text{input}} = \frac{110 \times 240}{4 \times 7{,}200} = \frac{26{,}400}{28{,}800} = 91.67\%$$

3-10   What instrument is used to measure voltage?
A voltmeter.

3-11   How is a voltmeter connected in a circuit (explain and illustrate)?
A voltmeter is always connected in shunt, or parallel, across the load or the source that you are measuring the voltage of. This is illustrated in Fig. 3-1.

Fig. 3-1.   Voltmeter connections in a circuit.

3-12   With what instrument do you measure current?
An ammeter.

3-13   How is an ammeter connected in a circuit (explain and illustrate)?
An ammeter is connected in series with the circuit being tested. This is illustrated in Fig. 3-2. Clamp-on ammeters, however, operate by measuring the magnetic field around conductors. These meters, though less accurate, are not wired into the circuit at all. Rather, they are simply clamped around the conductor whose current is to be measured.

Fig. 3-2.   Ammeter connections in a circuit.

3-14 Give two methods of measuring large currents. Explain why they must be used.

It is impractical to have a meter constructed that is capable of carrying the large currents. Therefore, when measuring large currents, in AC or DC circuits you use a shunt, which consists of a low resistance load connected in series with the load and also connected in parallel with a high-resistance meter. The meter then receives only a small fraction of the current passing through the load. This method is illustrated in Fig. 3-3.

Fig. 3-3. **One common method used to measure large current.**

Another method for measuring large currents in an AC circuit makes use of a current transformer (Fig. 3-4), which is described in Chapter 6. You may also use the wire that carries the load current as the primary of a current transformer in conjunction with a clip-on ammeter; the secondary of the transformer is incorporated in the meter, so the meter has no actual physical connection in the circuit (Fig. 3-4).

3-15 When using an ammeter, what precautions must be taken?

Know whether the current is AC or DC, and use the appropriate ammeter. Be sure that the rating of the meter is large enough for the current being measured; this step will prevent the meter from being damaged.

Fig. 3-4. **Measurement of large currents in an AC circuit.**

3-16 What is a wattmeter?

A wattmeter is an instrument that is designed to indicate directly the active power in an electric circuit. It consists of a coil connected in series with the circuit, such as in an ammeter, and a coil connected in parallel with the circuit, such as in a voltmeter. Both coils actuate the same meter, thereby giving the measurement of both the current and the voltage affecting one meter, which may be calibrated in watts, kilowatts, or megawatts.

3-17 How is a wattmeter connected in a circuit? Illustrate.

See Fig. 3-5.

Fig. 3-5. Wattmeter connections in a circuit.

3-18 Can the principles used in DC circuits be applied to all AC circuits?

The fundamental principles of DC circuits may also apply to AC circuits that are strictly resistive in nature, such as incandescent lighting and heating loads.

3-19 What causes inductive reactance?

Inductive reactance is caused by the opposition to the flow of alternating current by the inductance of the circuit.

3-20 Give some examples of equipment that cause inductive reactance.

Motors, transformers, choke coils, relay coils, ballasts.

3-21 When only inductive reactance is present in an AC circuit, what happens to the current in relation to the voltage?

The current is said to *lag* behind the voltage.

3-22 What is the reason for current lagging the voltage in an inductive circuit?

In an AC circuit, the current is continually changing its direction of flow (60 times a second in a 60-hertz circuit). Any change of current value is opposed by the inductance within the circuit involved.

3-23 Draw sine waves of voltage and current in a circuit containing only inductive reactance when 60-hertz AC is applied to the circuit.

See Fig. 3-6.

Fig. 3-6. Voltage and current sine waves in a circuit containing only inductive reactance, with a frequency of 60 hertz.

3-24 When only capacitive reactive is present in an AC circuit, what happens to the relationship that exists between the voltage and the current?

The current is said to *lead* the voltage.

3-25 Draw sine waves of current and voltage in a circuit containing only capacitive reactance when 60-hertz AC is applied to the circuit.

See Fig. 3-7.

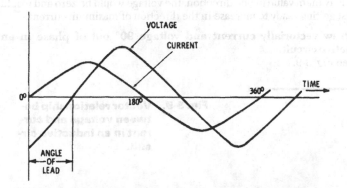

Fig. 3-7. Voltage and current sine waves in a circuit containing only capacitive reactance, with a frequency of 60 hertz.

3-26 In a purely resistive circuit with an applied 60-hertz AC voltage, what is the relationship between the current and voltage?

The current and voltage will be in phase, as illustrated in Fig. 3-8.

Fig. 3-8. Voltage-current relationship in a resistive circuit at a frequency of 60 hertz.

3-27 Is it possible to have a circuit with only inductive reactance?

Under ordinary circumstances, this is not possible, since all metal conductors have a certain amount of resistance. This is, however, possible for circuits that employ superconductors. At present, this can be done only in laboratory situations.

3-28 If it were possible to have only inductive reactance in an AC circuit, by what angle would the current lag the voltage?

The current would lag the voltage by 90°; that is, when the voltage was at its maximum value in one direction, the current would be zero and would be just getting ready to increase in the direction of maximum voltage.

3-29 If it were possible to have only capacitive reactance in an AC circuit, by what angle would the current lead the voltage?

The current would lead the voltage by 90°; that is, when the current was at its maximum value in one direction, the voltage would be zero and would be just getting ready to increase in the direction of maximum current.

3-30 Draw vectorially current and voltage 90° out of phase in an inductive circuit.

See Fig. 3-9.

Fig. 3-9. Vector relationship between voltage and current in an inductive circuit.

3-31 Draw vectorially current and voltage 90° out of phase in a capacitive circuit.

See Fig. 3-10.

3-32 What is meant by power factor?

Power factor is the phase displacement of current and voltage in an AC circuit. The cosine of the phase angle of displacement is the power factor; the cosine is multiplied by 100 and is expressed as a percentage. The cosine of 90° is 0; therefore, the power factor is 0%. If the angle of displacement were 60°, the cosine of which is 0.500, the power factor would be 50%. This is true whether the current leads or lags the voltage.

Fig. 3-10. Vector relationship between voltage and current in a capacitive circuit.

3-33 How is power expressed in DC circuits and AC circuits that are purely resistive in nature?

In DC circuits and AC circuits that contain only resistance,

$$P \text{ (watts)} = E \times I$$

$$P \text{ (kilowatts)} = \frac{E \times I}{1,000}$$

3-34 How is power expressed in AC circuits that contain inductive and/or capacitive reactance?

$$VA \text{ (volt-amperes)} = E \times I$$

$$KVA \text{ (kilovolt-amperes)} = \frac{E \times I}{1,000}$$

$$P \text{ (watts)} = E \times I \times \text{power factor}$$

3-35 In a 60-AC circuit, if the voltage is 120 volts, the current is 12 amperes, and the current lags the voltage by 60°, find (a) the power factor; (b) the power in volt-amperes (VA); (c) the power in watts.

(a) The cosine of 60° is 0.500; therefore, the power factor is 50%.
(b) 120 ×12 = 1,440 VA, which is called the *apparent* power.
(c) 120 × 12 × 0.5 = 720 watts, which is called the *true* power.

3-36 Show vectorially a 12-ampere line current lagging the voltage by 60°; indicate that the in-phase current is 50%, or 6 amperes.

See Fig. 3-11.

Fig. 3-11. A 12-ampere line current lagging the voltage by 60°; the in-phase current is 50% of the line current, or 6 amperes.

3-37 Why is a large power factor of great importance?

As can be seen in questions 3-35 and 3-36, there is an apparent power of 1,440 VA and a true power of 720 watts. There are also 12 amperes of line current and 6 amperes of in-phase, or effective, current. This means that all equipment from the source of supply to the power-consumption device must be capable of handling a current of 12 amperes, while actually the device is only utilizing a current of 6 amperes. A 50% power factor was used intentionally to make the results more pronounced. The I^2R loss is based on the 12-ampere current, while only 6 amperes are really effective.

3-38 How can power factor be measured or determined?

There are two easy methods: (1) by the combined use of a voltmeter, ammeter, and wattmeter (all AC instruments, of course), and (2) by the use of a power-factor meter.

3-39 How is the voltmeter-ammeter-wattmeter method used to determine the power factor?

The voltmeter, ammeter, and wattmeter are connected properly in the circuit. Then the readings of the three meters are simultaneously taken under the same load conditions. Finally, the following calculations are made:

$$\text{Power factor} = \frac{\text{true power (watts)}}{\text{apparent power } (E \times I)}$$

$$= \frac{w}{E \times I}$$

$$= \frac{kw}{KVA}$$

3-40 What is a power-factor meter?

A power-factor meter is a wattmeter calibrated to read the power factor directly instead of in watts. It is connected in the circuit in the same manner as a wattmeter.

3-41 In a circuit that contains an inductance, such as a motor, will the current lead or lag the voltage? What steps can be taken to correct the power factor?

The current will lag the voltage. The power factor can be corrected by adding capacitance to the circuit. This can be done either by introducing capacitors to the circuit, or by using synchronous motors and over-exciting their fields.

3-42 What is the ideal power-transmission condition, as far as the power factor is concerned?

Unity power factor, or a power factor of 1, which means that the current is in phase with the voltage.

3-43 Which is the best condition, a leading or lagging current?

They both have the same effect, so one is as good as the other. In an overall picture, there are usually more lagging currents than leading currents, since most of the loads used in the electrical field are of an inductive nature.

3-44 In an AC voltage or current, there are three values that are referred to. What are they?
The *maximum* value of current or voltage, the *effective* value of current or voltage, and the *average* value of current or voltage.

3-45 What is meant by the *maximum* AC voltage?
This is the maximum, or peak, voltage value of an AC sine wave.

3-46 What is meant by the *average* AC voltage?
This is the average of the voltages taken at all points on the sine wave.

3-47 What is meant by the *effective* AC voltage?
This is the value of the useful voltage that is indicated on a voltmeter. It is the voltage that is used in all normal calculations in electrical circuits.

3-48 What percentage of the peak voltage is the effective voltage?
70.7% of the peak value.

3-49 What percentage of the peak voltage is the average voltage?
63.7% of the peak voltage.

3-50 What is the effective voltage most commonly called?
Effective voltage is usually called rms voltage. (The term rms means root-mean-square.)

3-51 How could the rms voltage be arrived at vectorially?
An infinite number of lines could be drawn from the base line to the one-half amplitude value of the sine wave. These voltage values would then be squared; the sum of the squares would be averaged (added together and divided by the number of squares), and the square root of this figure would be the rms value of the voltage in question.

3-52 Does the peak voltage have to be considered?
Yes. In design, the peak voltage must be considered because this value is reached twice in every cycle.

3-53 Draw a sine wave showing the effective value and the peak value of a standard 120-volt AC line.
See Fig. 3-12.

Fig. 3-12. The effective and peak values of a standard 120-volt line.

CHAPTER 4

Lighting

4-1 What is a grounded (neutral) conductor?

A white or natural gray conductor, which indicates that it is the grounded circuit conductor.

4-2 May a grounded conductor ever be used as a current-carrying conductor? Explain.

It is permissible to use a white or natural gray conductor as a current-carrying conductor, if the conductor is permanently reidentified by painting or other effective means at each location where the conductors are visible and accessible (see *NEC*, Section 200-7, Exception No. 2). Note, however, that when used as we have just described, the white or gray conductor is removed from ground, and is no longer a grounded conductor.

4-3 Draw a diagram of the proper connections for a two-wire cable, one white and one black, to supply a light from a single-pole switch.

See Fig. 4-1.

Fig. 4-1.   A two-wire cable supplying a lamp from a single-pole switch.

97

4-4 When must the neutral, or grounded, conductor be provided with a switch?

A lamp or pump circuit on a gasoline dispensing island *must* provide both the neutral and the hot wires, which supply the light or the pump, with a switch (see *NEC*, Section 514-5).

4-5 Draw a three-way switch system and a lamp, using cable; show the colors and connections.

See Fig. 4-2.

Fig. 4-2.   A three-way switch system with a lamp.

4-6 Draw a circuit supplying a lamp that is controlled by two three-way switches and one four-way switch (use cable).

See Fig. 4-3.

Fig. 4-3.   A lamp circuit that is controlled by two three-way switches and one four-way switch.

4-7 Draw a master-control lighting system.
See Fig. 4-4.

4-8 Draw an electrolier switching circuit for controlling lights.
See Fig. 4-5.

4-9 Draw an electrolier switching circuit to control three sets of lamps.
See Fig. 4-6.

**A. One circuit with two groups
of lamps.**

**B. Two circuits with four groups
of lamps.**

Fig. 4-4. A master-control lighting system.

Fig. 4-5. An electrolier switch arrangement for control of lamp circuits.

4-10 What is a preheated-type fluorescent tube?

There are two external contacts at each end of a glass tube. Each set of contacts is connected to a specially treated tungsten filament. The inside of the tube is coated with a fluorescent powder. The type of powder used controls the color output of the tube. The tube is filled with an inert gas, such as argon, and a small drop of mercury to facilitate starting.

4-11 What kind of light does the fluorescent bulb produce within the bulb itself?

Ultraviolet light.

Fig. 4-6. An electroller switch arrangement for control of three groups of lamps. The sequence of operation is depicted diagrammatically.

4-12 Do the filaments stay lit during the operation of a fluorescent bulb? Explain.

No. They remain lit only at the start to vaporize the mercury; they are then shut off by the starter. Current is supplied to one contact on each end, thereby sustaining the mercury arc within the tube.

4-13 Draw a simple circuit of a fluorescent lamp and fixture.

See Fig. 4-7.

Fig. 4-7.   A fluorescent lamp and fixture circuit.

4-14 Explain the action of the glow-type starter.

When the switch is closed, the high resistance in the glow bulb of the starter produces heat, which causes the bimetallic U-shaped strip to close the contacts, thereby lighting the filaments in the fluorescent bulb. When the contacts close, the glow bulb cools and allows the contacts to open, thus disconnecting one side of the filaments. The arc within the fluorescent bulb is sustained without the filaments being heated, and the bulb lights.

4-15 Is the power factor of the circuit in Fig. 4-7 good or bad? Is the current leading or lagging the voltage?

The power factor is bad; the current is lagging the voltage because of the inductive reactance of the ballast.

4-16 How may the power factor of the circuit in Fig. 4-7 be corrected?

By the addition of capacitors in the ballast to counteract the lagging power factor.

4-17 What is the fixture in question 4-16 called?

A power-factor-corrected fixture.

4-18 What is a trigger-start fluorescent fixture?

This fixture uses a special trigger-start ballast that automatically preheats the filaments without the use of a starter.

4-19 Is a fluorescent fixture more efficient than an incandescent fixture?

Yes; however, no fixed efficiency can be quoted, since it varies with the size of the bulb, the ballast, etc. As a rule of thumb, a 40-watt fluorescent bulb is generally considered to put out about the same amount of light as a 100-watt incandescent bulb.

4-20 What is an instant-start Slimline fluorescent lamp?
This lamp has a single terminal at each end; the ballast is normally of the autotransformer type, which delivers a high voltage at the start and a normal voltage after the arc is established.

4-21 Does room temperature affect the operation of fluorescent lamps?
Yes, the normal fluorescent lamp is designed for operation at 50°F or higher.

4-22 If the operating temperature is expected to be below 50°F, what measures should be taken as far as using fluorescent lamps?
Use special lamps and starters that are designed for lower temperatures.

4-23 Do fluorescent lamps produce a stroboscopic effect? Explain.
Yes; this effect is due to the fact that at 60 hertz the current passes through zero 120 times a second.

4-24 How can the strobe effect be compensated for in a two-bulb system?
When the power factor is corrected, the capacitor is so connected that at the instant the current in one bulb is passing through zero, the current in the other bulb is not at zero, and the strobe effect goes unnoticed.

4-25 What is another way to minimize the strobe effect?
By using a higher frequency, such as 400 hertz.

4-26 Does a 400-hertz frequency have any other advantages?
Yes. Smaller and lighter ballasts can be used. This makes feasible the use of simple capacitance-type ballasts, which produce an overall gain in efficiency.

4-27 Does frequent starting and stopping of fluorescent lighting affect the bulb life?
The life of fluorescent bulbs is affected by starting and stopping. A bulb that is constantly left on will have a much longer life than one that is turned on and off often.

4-28 When lighting a football field with large incandescent bulbs, how can the light output of the bulbs be increased?
By using bulbs of a lower voltage rating than that of the source of supply. For example, by using a bulb that is rated at 105 volts on a 120-volt supply, the output can be increased roughly by 30%, but the life of the bulb will be cut by about 10%.

4-29 How is the light output of lamps rated?
In lumens.

4-30 How are light levels rated?
In foot-candles.

4-31 What is a foot-candle?
The amount of direct light emitted by one international candle on a square foot of surface, every part of which is one foot away.

CHAPTER 5

Branch Circuits and Feeders

5-1 Is the *NEC* a law?

It is not a law, but it is adopted into laws that are established by governmental agencies; it may be adopted in its entirety, in part, or with amendments.

5-2 Who has the responsibility for *Code* interpretations?

The administrative authority that has jurisdiction over endorsement of the *Code* has the responsibility for making *Code* interpretations (see *NEC*, Section 90-4). Usually, this is a local electrical inspector.

5-3 When are rules in the *NEC* mandatory and when are they advisory?

When the word "shall" is used, the rules are mandatory. Fine Print Notes (FPN) explain the intent of *Code* rules (see Section 110-1 of the *Code*).

5-4 According to the *NEC*, what are voltages?

Throughout the *Code*, the voltage considered shall be that at which the circuit operates. This is often expressed as, for example, 600 volts, nominal (see *NEC*, Section 220-2).

5-5 When wire gage or size is referred to, what wire gage is used?

The American Wire Gage (AWG) or in circular mils (CM).

5-6 When referring to conductors, what material is referred to?

Copper, unless otherwise specified. Where other materials are to be used, the wire sizes must be changed accordingly (see *NEC*, Section 110-5).

5-7   In what manner must the work be executed?
All electrical equipment must be installed in a neat and workmanlike manner (see *NEC*, Section 110-12).

5-8   May wooden plugs be used for mounting equipment in masonry, concrete, plaster, etc.?
No (see *NEC*, Section 110-13).

5-9   How may conductors be spliced or joined together?
They must be spliced or joined together by approved splicing devices or by brazing, welding, or soldering with a fusible metal or alloy (see *NEC*, Section 110-14).

5-10   When soldering, what precautions must be used?
All joints or splices must be electrically and mechanically secure before soldering and then soldered with a noncorrosive flux (see *NEC*, Section 110-14). This does not apply to conductors for grounding purposes; soldering is not allowed on these conductors.

5-11   How should splices or joints be insulated?
They must be covered with an insulation that is equivalent to the original conductor insulation (see *NEC*, Section 110-14).

5-12   Can an autotransformer be used on an ungrounded system?
The autotransformer must have a grounded conductor that is common to both primary and secondary circuits and is tied into a grounded conductor on the system supplying the autotransformer (see *NEC*, Section 210-9).
An autotransformer may be used to extend or add an individual branch circuit in an existing installation for equipment load without the connection to a similar grounded conductor when transforming from a nominal 208-volt supply or similarly from 240 volts to 208 volts.

5-13   On No.6 or smaller conductors, what means must be used for the identification of the grounded conductors?
Insulated conductors of No.6 or smaller, when used as grounded conductors, must be white or natural gray. On Type MI cable, the conductors must be distinctively marked at the terminal during installation (see *NEC*, Section 200-6).

5-14   How should conductors larger than No.6 be marked to indicate the grounded wire?
By use of white- or natural-gray-colored insulation or by identifying with a distinctive white marking at the terminals during installation (see *NEC*, Section 200-6).

5-15   How is the high-leg conductor of a 4-wire delta identified?
Where, on a 4-wire delta-connected secondary, the midpoint of one phase is grounded to supply lighting and similar loads, that phase conductor having the higher voltage to ground shall be orange in color or be indicated by tagging or other effective means at the point where a connection is to be made if the neutral conductor is present (see Section 215-8 of the *NEC*).

5-16 On a grounded system, which wire must be connected to the screw shell of a lampholder?
The grounded conductor [see *NEC*, Section 200-10(c)].

5-17 What will determine the classification of branch circuits?
The maximum permitted setting or rating of the overcurrent-protective device in the circuit (see *NEC*, Section 210-3).

5-18 What color coding is required on multiwire branch circuits?
The grounded conductor of a branch circuit shall be identified by a continuous white or natural gray color. Where conductors of different systems are installed in the same raceway, box, auxiliary gutter, or other types of enclosures, one system grounded conductor, if required, shall have an outer covering of white or natural gray. Each other system grounded conductor, if required, shall have an outer covering of white with an identifiable colored stripe (not green) running along the insulation or other and different means of identification. Ungrounded conductors of different voltages shall be a different color or identified by other means [see *NEC*, Section 210-5(a)].

5-19 How must a conductor that is only used for equipment and grounding purposes be identified?
By the use of a green color, or green with one or more yellow stripes, except if it is bare (see *NEC*, Section 210-5).

5-20 Can green-colored wire be used for circuit wires?
No. Green is intended for identification of equipment grounding conductors only (see *NEC*, Section 210-5).

5-21 What voltage is used between conductors that supply lampholders of the screw-shell type, receptacles, and appliances in dwellings?
Generally speaking, a voltage of 120 volts between conductors is considered the maximum. There are, however, some exceptions (see *NEC*, Section 210-6).

5-22 What are the exceptions to the ruling of 120 volts maximum between conductors?
Listed high-intensity discharge lighting fixtures with medium base lampholders, fixtures with mogul base lampholders, other types of lighting equipment (without screw shell lampholders), utilization equipment that is either permanently connected, or cord and plug connected [see *NEC*, Section 210-6(c)].

5-23 How shall you ground the grounding terminal of a grounding-type receptacle?
By the use of an equipment grounding conductor of green covered wire, green with one or more yellow stripes, or bare conductors. However, the armor of Type AC metal-clad cable, the sheath of MI or cable, or a metallic raceway is acceptable as a grounding means. Section 350-5 does not permit the general use of flexible metal conduit as a grounding means unless it and the connectors are listed [see *NEC*, Sections 210-7, 250-45, 250-57(a), 250-59, 250-91(b), and 350-5].

5-24 How can you ground the grounding terminal on a grounding-type receptacle on extensions to existing systems?

Run the grounding conductor to a grounded water pipe near the equipment (see *NEC*, Section 250-50).

5-25 What is the minimum size for branch-circuit conductors?

They cannot be smaller than No. 8 for ranges of 8 ¾ kw or more rating, and not smaller than No. 14 for other loads [see *NEC*, Section 210-19(a) and (b)].

5-26 What is the requirement concerning all receptacles on 15- and 20-ampere branch circuits?

All receptacles on 15- and 20-ampere branch circuits must be of the grounding type. A single receptacle installed on an individual branch circuit shall have a rating of not less than the rating of the branch circuit. [For complete details, see *NEC*, Section 210-7(a) and (b).]

5-27 What are the requirements for spacing receptacles in dwelling occupancies?

All receptacles in kitchens, family rooms, dining rooms, living rooms, parlors, libraries, dens, sun rooms, recreation rooms, and bedrooms must be installed so that no point along the wall space, measured horizontally, is more than 6 feet from a receptacle. This includes wall space that is 2 feet or wider and any space occupied by sliding panels on exterior walls. Sliding glass panels are excepted. At least one outlet shall be installed for the laundry. (See *NEC*, Section 210-52.) The wall space afforded by fitted room dividers, such as free-standing bar-type counters, shall be included in the 6-foot measurement.

In the kitchen and dining areas a receptacle outlet shall be installed at each counter space that is 12 inches wide or more. Countertop spacers separated by range tops, refrigerators, or sinks shall be considered as separate countertop space. Receptacles rendered inaccessible by the installation of appliances fastened in place or appliances occupying dedicated space will not be considered as the required outlets.

At least one wall receptacle outlet shall be installed in the bathroom adjacent to each basin location.

5-28 In residential occupancies, will ground-fault circuit interrupters be required?

Yes, for all 125-volt, 15- and 20-ampere receptacle outlets installed out-of-doors for residential occupancies and also for receptacle outlets in bathrooms, basements, kitchens, and garages (see *NEC*, Section 210-8).

5-29 A bedroom has one wall that contains a closet 8 feet in length, with sliding doors and a wall space of 2 feet; the bedroom door opens into the room and back against this wall space. Where are receptacles required on this wall?

One receptacle is required in the 2-foot space (Fig. 5-1). [See *NEC*, Section 210-52(a).]

Fig. 5-1. Receptacle requirements on a wall containing a closet with sliding doors.

5-30 Sketch a typical living room with sliding panels on the outside wall. Locate receptacles and show the room dimensions (Fig. 5-2).

Fig. 5-2. Receptacle requirements in a living room that has sliding doors on the outside wall.

5-31 A fastened-in-place appliance is located on a 15- or 20-ampere branch circuit on which there is a lighting fixture. What is the maximum rating that would be permitted on the appliance?

50% of the branch circuit rating [see *NEC*, Section 210-23(a)].

5-32 A cord-and-plug-connected appliance is used on a 15- or 20-ampere branch circuit. What is the maximum rating permitted for the appliance?

80% of the branch circuit rating [see *NEC*, Section 210-23(a)].

5-33 What is the smallest size wire permissible for a feeder circuit?
No. 10 wire [see *NEC*, Section 215-2(a)].

5-34 What are the permissible voltage drops allowable on feeders and branch circuits?
On feeders, not more than 3% for power, heating, and lighting loads or combinations thereof. A total maximum voltage drop not to exceed 5% for conductors and for combinations of feeders and branch circuits [see *NEC*, Section 215-2(b), FPN].

5-35 What is the basis for figuring the general lighting loads in occupancies?
They are figured on a volt-amperes-per-square-foot basis, using *NEC*, Table 220-3(b) can be used to determine the unit load per square foot in volt amperes for the types of occupancies listed.

5-36 What measurements are used to determine the number of watts per square foot?
The outside dimensions of the building and the number of floors, not including open porches or garage [see *NEC*, Section 220-3(b)].

5-37 What is the unit load per square foot (in volt-amperes) for a hospital?
Two volt-amperes [see *NEC*, Table 220-3(b)].

5-38 What is the unit load per square foot (in volt-amperes) for a school?
Three volt-amperes [see *NEC*, Table 220-3(b)].

5-39 How many volt-amperes per square foot are required to be included for general purpose receptacle outlets when a set of plans does not show their locations in an office building?
One volt-ampere per square foot [see *NEC*, Table 220-3(b)** note].

5-40 What is the unit load per square foot (in volt-amperes) for a warehouse used for storage?
¼ volt-amperes [see *NEC*, Table 220-3(b)].

5-41 What voltages are used for purposes of computing feeder and branch circuit loads when other voltages are not specified?
120, 120/240, 208Y/120, 240, 480Y/277, 480, and 600 volts (see *NEC*, Section 220-2).

5-42 How are continuous and noncontinuous loads for branch circuit ratings calculated?
Branch circuits must be rated no less than the noncontinuous load, plus at least 125% of the continuous load [see *NEC*, Section 220-3(a)].

5-43 Are unfinished basements of dwellings used in figuring watts per square foot?
Yes, if adaptable for future use; if these spaces are not adaptable, they are not used [see *NEC*, Section 220-3(b)].

5-44 What loads are used in figuring outlets for other than general illumination?

Outlets supplying specific loads and appliances must use the ampere rating of the appliance. Outlets supplying heavy-duty lampholders must use 600 volt-amperes; calculations for other outlets must use 180 volt-amperes [see *NEC*, Section 220-3(c)].

5-45 What load must be figured for show-window lighting?
Not less than 200 volt-amperes for each linear foot measured horizontally along the base (see *NEC*, Section 220-12).

5-46 What are the receptacle requirements in dwelling occupancies for kitchen, family room, laundry, pantry, dining room, and breakfast room?
There shall be a minimum of two 20-ampere, small-appliance circuits for the kitchen, family room, pantry, dining room, and breakfast room. Also, a minimum of one 20-ampere circuit for the laundry [see *NEC*, Section 220-4(b) and (c)].

5-47 What is the unit load per square foot (in volt-amperes) for a store?
Three volt-amperes [see *NEC*, Table 220-3(b)].

5-48 Outlets for heavy-duty lampholders are to be based on a load of how many volt-amperes?
600 [see *NEC*, Section 220-3(c)(4)].

5-49 How is the load for a household electric clothes dryer computed?
See *NEC*, Section 220-18 and Section 220-3(c)(1), Exception No. 5.

5-50 Are demand factors permitted in determining feeder loads?
Yes; they are given in *NEC*, Table 220-11.

5-51 What size feeders must be installed in dwelling occupancies?
The computed load of a feeder must not be less than the sum of all branch circuit loads supplied by the feeder. The demand factors may be used in the calculation of feeder sizes.

5-52 When figuring the neutral load to electric ranges, what is the maximum unbalanced load considered to be?
The maximum unbalanced load for electric ranges is considered to be 70% of the load on the ungrounded conductors (see *NEC*, Section 220-22).

5-53 How is the neutral load on a 5-wire, 2-phase system determined?
It is figured at 140% of the load on the ungrounded conductors (see *NEC*, Section 220-22).

5-54 How may the neutral-feeder load on a 3-wire DC or single-phase 3-wire AC system be determined?
The 70% demand factor may be used on range loads, and a further demand factor of 70% may be used on that portion of the unbalanced load in excess of 200 amperes (see *NEC*, Section 220-22).

5-55 How do you calculate the unbalanced load on a 4-wire, 3-phase system?
The 70% demand factor may be used on range loads, and a further demand factor of 70% may be used on that portion of the unbalanced load in

excess of 200 amperes (see *NEC*, Section 220-22).

5-56 Can you make a reduction on the neutral feeder load where discharge lighting, data processing, or similar equipment is involved?

No reduction can be made on the neutral capacity for that portion of the load which consists of electric discharge lighting, data processing or similar equipment. The load on the neutral feeder must be taken at 100% of the ungrounded conductors (see *NEC*, Section 220-22).

5-57 Why do discharge lighting loads require no reduction in neutral feeder capacity?

Because of the effect of the third harmonic on the current value in the neutral feeder.

5-58 Are demand factors applicable to electric ranges?

Yes. *NEC*, Table 220-19 may be used in determining demand factors for elec electric ranges.

5-59 Can demand factors be used on electric-clothes-dryer loads in the same manner as they are used on electric ranges?

Yes (see *NEC*, Table 220-18).

5-60 Are feeder demand factors permitted for commercial ranges and other commercial kitchen equipment?

Yes (see *NEC*, Table 220-20).

5-61 There are 2,500 square feet of floor area in a house which contains an electric range rated as 12 kW and an electric dryer rated at 5,000 watts. Calculate: (a) the general lighting required; (b) the minimum number and sizes of branch circuits required; (c) the minimum, size of feeders (service conductors) required.

(a)   2,500 sq. ft. at 3 volt-amperes per sq. ft. equals 7,500 watts.

(b)   7,500 watts divided by 120 volts equals 62.5 amperes; this would require a minimum of five 15-ampere circuits with a minimum of No. 14 wire, or a minimum of four 20-ampere circuits with a minimum of No. 12 wire.

Small-appliance load: A minimum of two small-appliance circuits [see *NEC*, Section 220-4(b)] of 1,500 volt-amperes each; these will require 20-ampere 2-wire circuits with No. 12 wire.

A 12-kW range has a demand of 8 kW, according to *NEC*, Table 220-19. 8,000 watts divided by 240 volts equals 33 amperes; therefore, a minimum of No. 8 wire with a 40-ampere service would be required, although good practice would indicate the use of No. 6 wire with a 50-ampere circuit breaker.

The dryer: 5,000 watts divided by 240 volts equals 21 amperes; therefore, you would use No. 10 wire with a 30-ampere circuit.

Laundry circuit: 1,500 watts.

(c)   Minimum size of service.

General lighting	7,500 volt-amperes
Small appliance load	3,000 volt-amperes
Laundry circuit	1,500 volt-amperes
	12,000 volt-amperes
3,000 volt-amperes @ 100%	3,000 volt-amperes
9,000 volt-amperes @ 35%	3,150 volt-amperes
Net computed (without range and dryer)	6,150 volt-amperes
Range load	8,000 volt-amperes
Dryer load	5,000 volt-amperes
	13,000 volt-amperes
Net computed (with range and dryer)	19,150 volt amperes

Loads of over 10 kW must have a minimum of 100-ampere service; therefore, the minimum service will be 100 amperes. We can use No. 4 copper THW conductors with a 100-ampere main disconnect (see *NEC*, Note 3 to Tables 310-16 through 310-31).

5-62 What are the demand factors for nondwelling receptacle loads?
See *NEC*, Table 220-13.

5-63 How are motor loads computed?
In accordance with Sections 430-24, 430-25, and 430-26 (see *NEC*, Section 220-14).

5-64 How are fixed electric space heating loads computed?
At 100% of the total connected load but in no case is a feeder load current rating to be less than the rating of the largest branch-circuit supplied (see *NEC*, Section 220-15).

5-65 What percentage is permitted to be applied as a demand factor to the nameplate rating load for four or more appliances fastened in place and served by the same feeder in a one-family, two-family, or multifamily dwelling?
Seventy-five percent for all appliances except electric ranges, clothes dryers, space heating equipment, or air-conditioning equipment (see *NEC*, Section 220-17).

5-66 The demand factor for 5 household electric clothes dryers is calculated at what percentage?
80 percent (see *NEC*, Table 220-18).

5-67 Where two or more single-phase ranges are supplied by a 3-phase, 4-wire feeder, how is the total load computed?
On the basis of twice the maximum number connected between any two phases (see *NEC*, Section 220-19).

5-68 What is the maximum demand for two household electric ranges, wall-mounted ovens, counter-mounted cooking units, and other household cooking appliances rated at over 1 ¾ kW?
11 kW for ratings not over 12 kW (see *NEC*, Section 220-19).

5-69 What computations are required for commercial cooking equipment?

See *NEC*, Table 220-20.

5-70 When computing the total load of a feeder for an electric heating system and an air-conditioning system that will not be used simultaneously must both loads be added?

No. The smaller of the two can be omitted (see *NEC*, Section 220-21).

5-71 A store building is to be wired for general illumination and showwindow illumination. The store is 40 feet by 75 feet, with 30 linear feet of show window. With a density of illumination of 3 watts per square foot for the store and 200 volt-amperes per linear foot for the showwindow, calculate: (a) the general store load; (b) the minimum number of branch circuits required and sizes of wire; (c) the minimum size of feeders (or service conductors) required.

(a)   General lighting load.

3,000 sq. ft. at 3 volt-amperes per sq. ft. equals 9,000 volt-amperes. However, this load will be required most of the time, so we must multiply 1.25 times 9,000, or 11,250 volt-amperes.

Show window: 30 linear feet at 200 volt-amperes per foot equals 6,000 volt-amperes. Therefore, the general store load is 11,250 volt-amperes.

(b)   Minimum number of branch circuits and wire sizes.

11,250 volt-amperes divided by 240 volts equals 47 amperes. For three-wire service, this current will require four 15-ampere circuits using No. 14 wire (minimum) or three 20-ampere circuits using No. 12 wire (minimum).

Show window: 6,000 volt-amperes divided by 240 volts equals 25 amperes. Therefore, two 15-ampere circuits with No. 14 wire or two 20-ampere circuits with No. 12 wire will be required.

(c)   Minimum size service conductors required.

The ampere load for three-wire service would be 47 amperes plus 25 amperes equals 72 amperes. Therefore, a 100-amp service would be used with No. 3 THW conductors in 1¼-inch conduit, or No. 3 THHN conductors in 1-inch conduit.

If the service is 120 volts instead of 120/240 volts (which is highly improbable because the utility serving would require the 120/240 volts), then we would have to use 120 volts when finding the current. These currents would then be doubled; entrance-wire capacities would double as well as the main disconnect on the service entrance.

5-72 Determine the general lighting and appliance load requirements for a multifamily dwelling (apartment house). There are 40 apartments, each with a total of 800 square feet. There are two banks of meters of 20 each, and individual subfeeders to each apartment.

Twenty apartments have electric ranges; these apartments (with ranges) are evenly divided, 10 on each meter bank; the ranges are 9 kW each. The service is 120/240 volts. Make complete calculations of what will be required, from the service entrance on.

Computed load for each apartment (see *NEC*, Articles 220).

> General lighting load:
> 800 sq. ft. @ 3-volt amperes per sq. ft. .... 2,400 volt-amperes
> Small-appliance load .... 3,000 volt-amperes
> Electric range .... 3,000 volt-amperes

Minimum number of branch circuits required for each apartment (see *NEC*, Section 220-4).

> 2,400 volt-amperes divided by 120 volts equals 20 amperes. This current will require two 15-ampere circuits using No. 14 wire or two 20-ampere circuits using No. 12 wire.

Small-appliance load:
> Two 20-ampere circuits using No. 12 wire [see *NEC*, Section 220-4(b)].

Range circuit:
> 8,000 watts divided by 240 equals 33 amperes. A circuit of two No. 8 wires or one No. 10 wire is required [see *NEC*, Section 210-19 (b)].

Minimum size subfeeder required for each apartment (see *NEC*, Section 215-2).

Computed load:

> General lighting load .... 2,400 volt-amperes
> Small-appliance load (two 20-ampere
> circuits) .... 3,000 volt-amperes
> Total computed load
> (without ranges) .... 5,400 volt-amperes

Application of demand factor:

> 3,000 watts @ 100% .... 3,000 volt-amperes
> 2,400 watts @ 35% .... 840 volt amperes

> Net computed load
> (without ranges) .... 3,840 volt-amperes
> Range load .... 8,000 volt-amperes

> Net computed load (with ranges) .... 11,840 volt-amperes

For 120/240 volt, 3-wire system (without ranges):
> Net computed load:
> 3,840 volt-amperes divided by 240 volts equals 16 amperes.
> Minimum feeder size could be No. 10 wire with a two-pole, 30-ampere circuit breaker.

For 120/240 volt, 3-wire system (with ranges):
> Net computed load:
> 11,840 volt-amperes divided by 240 volts equals 49 amperes.
> Minimum feeder size would be No. 6 wire with two 60-ampere fuses or No. 4 wire with a two-pole, 70-ampere circuit breaker.

Neutral subfeeder:

Lighting and small-appliance load .... 3,840 volt-amperes
Range load (8,000 watts @ 70%)
(see *NEC*, Section 220-22) .... 5,600 volt-amperes

Net computed load (neutral) .... 9,440 volt-amperes
9,400 watts divided by 240 volts equals 41 amperes. Size of neutral subfeeder would be No.6 wire. There would also be two main disconnects needed ahead of the meters.

Minimum size feeders required from service equipment to meter bank (for 20 apartments—10 with ranges).
Total computed load:
Lighting and small-appliance load—20 times 5,400 volt-amperes equals 108,000 volt-amperes.
Application of demand factor:
3,000 volt-amperes @ 100% .... 3,000 volt-amperes
105,000 volt-amperes @ 35% .... 36,750 volt-amperes

Net computed lighting and
small-appliance load .... 39,750 volt-amperes
Range load (10 ranges, less than
12 kW .... 25,000 volt-amperes

Net computed load (with ranges) .... 64,750 volt-amperes
For 120/240 volt, 3-wire system:
Net computed load: 64,750 divided by 240 equals 270 amperes. Size of each ungrounded feeder to each meter bank would be 500,000 CM.
Neutral feeder:
Lighting and small-appliance load .... 39,750 volt-amperes
Range load (25,000 watts @ 70%) .... 17,500 volt-amperes

Computed load (neutral) .... 57,250 volt-amperes

57,250 volt-amperes divided by 240 volt equals 239 amperes.
Further demand factor:
200 amperes @ 100% .... 200 amperes
39 amperes @ 70% .... 27 amperes

Net computed load (neutral) .... 227 amperes

Minimum size main feeder (service conductors) required (for 40 apartments—20 with ranges).
Total computed load:
Lighting and small-appliance load—40 times 5,400 volt-amperes ....
equals 216,000 volt-amperes.

Application of demand factor:
3,000 volt-amperes @ 100% .... 3,000 volt-amperes
117,000 volt-amperes @ 35% .... 40,950 volt-amperes
96,000 volt-amperes @ 25% .... 24,000 volt-amperes

Net computed lighting and small-appliance
load .... 67,950 volt-amperes
Range load (20 ranges, less than 12 kW .... 35,000 volt-amperes

Net computed load .... 102,950 volt-amperes

For 120/240 volt, 3-wire system:
Net computed load—102,950 volt-amperes divided by 240 volts equals
429 amperes.
Size of each ungrounded main feeder would be 1,000,000 CM.

Neutral feeder:
Lighting and small appliance load .... 67,950 volt-amperes
Range load (35,000 volt-amperes @ 70%) .... 24,500 volt-amperes

Computed load (neutral) .... 92,450 volt-amperes
92,450 volt-amperes divided by 240 volts equals 385 amperes.

Further demand factor:
200 amperes @ 100% .... 200 amperes
185 amperes @ 70% .... 130 amperes

Net computed load (neutral) .... 330 amperes

Size of neutral main feeder would be 6,000,000 CM.

5-73 There is to be a current of 100 amperes per phase on a 4-wire
120/208-volt wye system. What size neutral would we need? The phase
wires are No. 2 THHN.

Since in this case neutrals are not allowed to be derated, and must be the
same size as the phase conductors, we would need to use a No. 2 THHN conductor.

5-74 Is there a demand factor for feeders and service entrance conductors for multifamily dwellings?

Yes, *NEC*, Table 220-32.

CHAPTER 6

Transformer Principles and Connections

6-1 What is induction?

The process by which one conductor produces, or induces, a voltage in another conductor, even though there is no mechanical coupling between the two conductors.

6-2 What factors affect the amount of induced electromotive force (emf) in a transformer?

The strength of the magnetic field, the speed at which the conductors are cut by the magnetic field, and the number of turns of wire being cut by the magnetic field

6-3 What is inductance?

The property of a coil in a circuit to oppose any change of existing current flow.

6-4 What is self-inductance?

The inducing of an emf within the circuit itself, caused by any change of current within that circuit. This induced emf is always in a direction opposite to the applied emf, thus causing opposition to any change in current within the circuit itself.

6-5 What is mutual inductance?

The linkage of flux between two coils or conductors, caused by the current flowing within one or both of the coils or conductors.

119

6-6 Draw a diagram of two coils, such as the coils of a transformer winding, and indicate the self-inductance and the mutual inductance.

Self-inductance is produced within the primary coil and mutual inductance exists between the two transformer coils, as shown in Fig. 6-1.

Fig. 6-1. Self-inductance and mutual inductance in the coils of a transformer.

6-7 Name several methods by which an emf may be generated.

By conductors being cut by a magnetic field (generators), by chemical action (batteries), by heat (thermocouples), by crystal vibration (phono cartridges), and by friction.

6-8 What is direct current (DC)?

Current that flows in one direction only.

6-9 What is alternating current (AC)?

Current that continually reverses its direction of flow.

6-10 What is pulsating direct current?

An unidirectional current that changes its value at regular or irregular intervals.

6-11 What is a cycle?

One complete alternation, or reversal, of alternating current. The wave rises from zero to maximum in one direction, falls back to zero, then rises to maximum in the opposite direction, and finally falls back to zero again.

6-12 What always surrounds a conductor when a current flows through it?

A magnetic field.

6-13 What is the phase relation between the three phases of a three-phase circuit?

They are 120 electrical degrees apart.

6-14 Draw sine waves for three-phase voltage; show polarity, time, and phase angle (in degrees).

See Fig. 6-2.

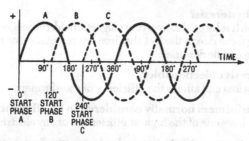

Fig. 6-2. Three-phase voltage sine waves.

6-15 What is the phase relation between phases in two-phase circuit?
They are 90 electrical degrees apart.

6-16 What is a transformer?
A device that transforms electrical energy from one or more circuits to one or more other circuits at the same frequency but usually at a different voltage and current. It consists of a core of soft-iron laminations, surrounded by coils of copper-insulated wire.

6-17 There are two basic types of transformers. What are they?
The isolation type—the two windings are physically isolated and electrically insulated from each other; and the auto-transformer type—there is only one coil with a tap or taps taken off it to secure other voltages (the primary is part of the secondary and the secondary is part of the primary).

6-18 What is an oil-immersed transformer?
The core and coils are immersed in a high-grade mineral oil, which has high dielectric qualities.

6-19 Why is oil used in a transformer?
To increase the dielectric strength of the insulation, to keep down the possibility of arcing between coils, and to dissipate heat to the outer case so that the transformer may carry heavier loads without excessive overheating.

6-20 What is an air-core transformer?
A transformer that does not contain oil or other dielectric compositions but is insulated entirely by the winding insulations and air.

6-21 What are eddy currents?
Circulating currents induced in conductive materials (the iron cores of transformers or coils) by varying magnetic fields.

6-22 Are eddy currents objectionable?
Yes; they represent a loss in energy and also cause overheating.

6-23 What means can be taken to keep eddy currents at a minimum?
The iron used in the core of an alternation-current transformer is laminated, or built up of thin sheets or strips of iron, so that eddy currents will circulate only in limited areas.

6-24 What is hysteresis?

When iron is subjected to a varying magnetic field, the magnetism lags the magnetizing force due to the fact that iron has reluctance, or resistance, to changes in magnetic densities.

6-25 Is hysteresis objectionable?

Yes; it is a loss and affects the efficiency of transformers.

6-26 Are transformers normally considered to be efficient devices?

Yes; they have one of the highest efficiencies of any electrical device.

6-27 What factors constitute the major losses produced in transformers?

Power loss of the copper (I^2R losses), eddy currents, and hysteresis losses.

6-28 Is there a definite relationship between the number of turns and voltages in transformers?

Yes; the voltage varies in exact proportion to the number of turns connected in series in each winding.

6-29 Give an illustration of the relationship between the voltages and the turns ratio in a transformer.

If the high-voltage winding of a transformer has 1,000 turns and a potential of 2,400 volts is applied across it, the low-voltage winding of 100 turns will have 240 volts induced across it; this is illustrated in Fig. 6-3.

Fig. 6-3. Relationship between voltages and the turns ratio in a transformer.

6-30 What is the difference between the primary and the secondary of a transformer?

The primary of the transformer is the input side of the transformer and the secondary is the output side of the transformer. On a step-down transformer, the high-voltage side is the primary and the low-voltage side is the secondary; on a step-up transformer, the opposite is true (Fig. 6-4).

Fig. 6-4. The primary and secondary windings of a step-down and a step-up transformer.

6-31 Ordinarily, what is the phase relationship between the primary and secondary voltages of a transformer?

They are 180° out of phase.

6-32 Is it possible to have the primary and secondary of a transformer in phase?

Yes; by changing the connections on one side of the transformer.

6-33 How are the leads of a transformer marked, according to ANSI (American National Standards Institute)?

The high side of the transformer is marked H_1, H_2, etc. The low side of the transformer is marked X_1, X_2, etc.

6-34 What is the purpose of the markings on transformer leads?

They are there for standardization, so that transformer polarities may be recognizable for any type of use.

6-35 Draw a diagram of a transformer with additive polarity, using ANSI markings.

See Fig. 6-5.

ADDITIVE POLARITY

Fig. 6-5. A transformer with additive polarity.

6-36 Draw a diagram of a transformer with subtractive polarity, using ANSI markings.

See Fig. 6-6.

Fig. 6-6. A transformer with subtractive polarity.

SUBTRACTIVE POLARITY

6-37 If a transformer is not marked, how could you test it for polarity?
Connect the transformer as shown in Fig. 6-7. If it has subtractive polarity, V will be less than the voltage of the power source; if it has additive polarity, V will be greater than the voltage of the power source.

Fig. 6-7. Testing a transformer for polarity.

6-38 What is a split-coil transformer?
A transformer that has the coils on the low or high side in separate windings, so that they can be connected in series or parallel for higher or lower voltages, as desired.

6-39 Draw a diagram of a split-coil transformer with the low side having split coils for dual voltages; draw an additive polarity transformer, and mark the terminals with ANSI markings. Show the voltages that you use.
See Fig. 6-8.

6-40 Draw a diagram of an autotransformer.
See Fig. 6-9.

Fig. 6-8. A split-coil transformer and an additive-polarity transformer.

Fig. 6-9. An autotransformer.

6-41 Where may autotransformers be used?

(a) Where the system being supplied contains an identified grounded conductor that is solidly connected to a similar identified grounded conductor of the system supplying the autotransformer (see *NEC*, Sections 210-9 and 450-4); (b) where an autotransformer is used for starting or controlling an induction motor [see *NEC*, Section 430-82(b)]; (c) where an autotransformer is used as a dimmer, such as in theaters [see *NEC*, Section 520-25(c)]; (d) as part of a ballast for supplying lighting units (see *NEC*, Section 410-78). For voltage bucking and boosting see *NEC*, Section 210-9, exception 1.

6-42 What is the relationship between the current and voltage in the high side of a transformer and the current and voltage in the low side of a transformer? Draw a diagram showing this relationship.

With respect to the turns ratio, the current in one side of a transformer is inversely proportional to the current in the other side, while the voltage across one side of a transformer is directly proportional to the voltage across the other side. These are illustrated in Fig. 6-10.

Fig. 6-10. Current-voltage relationship between the high side and the low side of a transformer.

6-43 When an induction coil is connected in a DC circuit, as in Fig. 6-11, what happens when the switch is closed? When the switch is opened?

When the switch is closed, the current slowly rises to a maximum (point *A* in this example). The retarding of current flow is due to self-inductance. After reaching the maximum at point *A*, the current will remain constant until the switch is opened (point *B*). When the switch is opened, the flux around the coil collapses, thereby causing an opposition to the current discharge; however, this discharge-time collapse is extremely short when compared to the charging time. The discharge causes a high voltage to be applied across the switch, which tends to sustain an arc; this voltage often reaches large values. The principles of this type of circuit have many applications, such as ignition coils and flyback transformers.

6-44 Draw a schematic diagram of the high-side windings of three single-phase transformers connected in a delta arrangement. Show the ANSI markings.

Note that in a delta arrangement, as shown in Fig. 6-12, H_1, is connected to H_2 of the next transformer, and so on.

Fig. 6-11. An induction coil in a DC circuit with the switch being opened and closed.

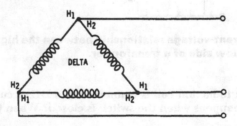

Fig. 6-12. Three single-phase transformer windings connected in a delta arrangement.

6-45 Draw a schematic diagram of the high-side windings of three single-phase transformers connected in a wye (Y) arrangement. Show the ANSI markings.

Note that in a wye arrangement, as shown in Fig. 6-13, all of the H_2's are connected in common, and the H_1's each supply one phase wire.

Fig. 6-13. Three single-phase transformer windings connected in a wye arrangement.

6-46 If the polarity of one transformer is reversed on a delta bank of single-phase transformers that are connected for three-phase operation, what would be the result?

Instead of zero voltage on the tie-in point, there would be a voltage that is twice the proper value.

6-47 Show with diagrams how you would test to be certain that the polarities on a delta bank of three single-phase transformers are correct.

See Fig. 6-14.

Fig. 6-14. Polarity tests on a delta bank of three single-phase transformer windings.

6-48 A bank of three single-phase transformers are connected in a delta; each transformer delivers 240 volts at 10 amperes. What are the line voltages and line currents?

The line voltages are each equal to 240 volts; however, the line current in each phase would be the current of each transformer times 1.732 (the square root of 3), or 17.32 amperes.

6-49 Draw a schematic diagram showing all the currents and voltages on a bank of three single-phase transformers that are connected in a delta arrangement. Assume a voltage across each transformer of 240 volts and a current through each transformer of 10 amperes.

See Fig. 6-15.

LINE CURRENT IS √3 X COIL
CURRENT IN A DELTA
CONNECTION

Fig. 6-15. Current and voltage values on a delta bank of three single-phase transformer windings. A voltage of 240 volts and a current of 10 amperes are assumed.

6-50 Draw a schematic diagram for a bank of three single-phase transformers that are connected in a wye. Show all voltages and currents; assume a voltage across each transformer of 120 volts and a current in each winding of 10 amperes.

See Fig. 6-16.

6-51 Draw a bank of three single-phase transformers that are connected in a delta-delta bank with one side connected to 2,400 volts, three-phase, and with the other side delivering 240 volts, three-phase. Show voltages and ANSI markings on all transformers.

See Fig. 6-17.

6-52 Is it possible to connect two single-phase transformers to secure a three-phase output from a three-phase input?

Yes; they would have to be connected in an open delta.

WYE CONNECTION
LINE VOLTS = 1.732 × COIL VOLTS
COIL AMPS = LINE AMPS

Fig. 6-16. Current and voltage values on a wye bank of three single-phase transformer windings. A voltage of 120 volts and a current of 10 amperes are assumed.

6-53 If you have a bank of three single-phase transformers that are connected in a closed delta arrangement, and one transformer burns up, how would you continue operation on the remaining two transformers?

By merely disconnecting the leads to the disabled transformer.

6-54 When you use a bank of two single-phase transformers in an open delta arrangement, do they supply their full output rating?

No. Each transformer is only capable of supplying 86.6% of its output rating.

6-55 If you have a bank of three single-phase transformers, each with a 10-kVA rating, that are connected in a closed delta arrangement, you would have a capacity of 30 kVA. If one transformer is taken out of the bank, what would be the output capacity of the remaining 10-kVA transformers?

Each transformer would deliver 8.66 kVA, or you would have a bank capacity of 17.32 kVA.

6-56 Draw a schematic diagram of two transformers that are connected in an open delta arrangement. Show transformer voltages and the three-phase voltages.

See Fig. 6-18.

Fig. 6-17. Three single-phase transformers connected in a delta-delta bank; the high side is connected to 2400 volts, three-phase, and the low side delivers 240 volts, three phase.

Fig. 6-18. Two transformers in an open delta arrangement.

6-57 Draw a schematic diagram of three transformers that are connected in a delta arrangement on both sides, fed from a 2,400-volt source on the high side, and connected for a 240-volt, three-phase, and a 120/240-volt, single-phase output on the low side. Show all voltages.

See Fig. 6-19.

6-58 Draw a schematic diagram of three single-phase transformers that are connected in a wye-wye arrangement. Show the neutral on both high and low sides.

See Fig. 6-20.

DELTA-DELTA
HIGH VOLTAGE – 2400V, 3 Ø
LOW VOLTAGE – 240V, 3 Ø
and
120V/240V – 1 Ø WITH
208V FROM NEUTRAL
TO WILD LEG

Fig. 6-19. Three transformer connected in a delta-delta bank. The high side is connected to 2400 volts, three-phase, and the low side delivers 240 volts, three-phase, and 120/240 volts, single-phase.

6-59 Draw a schematic diagram of three single-phase transformers that are connected in a wye arrangement on the high side and a delta arrangement on the low side.

See Fig. 6-21.

6-60 What are instrument transformers?

In the measurement of current, voltage, or kilowatt-hours on systems with high voltage or high current, it is necessary to use a device known as an instrument transformer, which reproduces in its secondary circuit the primary current or voltage while preserving the phase relationship to measure or record at lower voltages or lower amperages, and then to use a constant to multiply the readings to obtain the actual values of voltage or current. Current transformers (CTs) are used to measure the current, and potential transformers (PTs) are used to register the potential.

6-61 Describe a potential transformer.

A potential transformer is built like the ordinary isolation transformer, except that extra precautions are taken to ensure that the winding ratios are exact. Also, the primary winding is connected in parallel with the circuit to be measured.

Fig. 6-20. Three single-phase transformer windings connected in a wye-wye arrangement.

Fig. 6-21. Three single-phase transformer windings connected in a wye arrangement on the high side and a delta arrangement on the low side.

6-62 Describe a current transformer.

A current transformer has a primary of a few turns of heavy conductor capable of carrying the total current, and the secondary consists of a number of turns of smaller wire. The primary winding is connected in series with the circuit carrying the current that is to be measured.

6-63 How are current transformers rated?

They are rated at 50 to 5, 100 to 5, etc. The first number is the total current that the transformer is supposed to handle, and the second figure is the current on the secondary when the full-load current is flowing through the primary. For example, a 50-to-5 rating would have a multiplier of 10 (K=10).

6-64 What precautions must be taken when working with current transformers? Why?

The secondary must never be opened when the primary circuit is energized. If it is necessary to disconnect an instrument while the circuit is energized, the secondary must be short-circuited. If the secondary is opened while the circuit is energized, the potential on the secondary may reach dangerously high values. By short-circuiting the secondary, damage is avoided and the voltage on the secondary is kept within safe limits.

6-65 Draw a schematic diagram of a current transformer.

As shown in Fig. 6-22, the primary consists of a single conductor; it may be a single conductor or only a few turns.

Fig. 6-22. A current transformer.

6-66 What is a booster transformer?

A transformer arrangement that is often used toward the end of a power line in order to raise the voltage to its desired value.

6-67 May an ordinary transformer be used as a booster transformer?

Yes.

6-68 When connection an ordinary transformer as a booster transformer, what important factors must be considered?

The high side of the transformer must be able to handle the approximate voltage of the line; the low side must have a voltage of approximately the value you wish to boost the line voltage, and the low side must also have a current capacity that is sufficient to carry the line current.

6-69 What special precaution must be taken when using a booster transformer?

There must be no fusing in the high side, or primary. Since the booster transformer is similar to a current transformer, if the fuse should blow, an extremely high voltage could be built up on the secondary side.

6-70 Draw a schematic diagram of an ordinary transformer that is connected as a booster transformer.

See Fig. 6-23.

Fig. 6-23. A voltage-booster transformer.

6-71 What is an induction regulator?

This device is similar to a booster transformer. It has a primary and a secondary winding which are wound on separate cores. The primary can be moved in either direction; this is usually done by an electric motor. In turning, the primary bucks or boosts the line voltage, as required. The amount of bucking or boosting is anticipated by the current being drawn by the line.

6-72 Draw a schematic diagram showing how an induction regulator is connected into the line.

See Fig. 6-24.

Fig. 6-24. The connection of an induction regulator to the line.

6-73 What is a three-phase transformer?

A transformer that is the equivalent of three single-phase transformers, which are all wound on one core and enclosed within one common case.

6-74 When connecting transformers in parallel, what factors must be taken into consideration?

Their electrical characteristics, such as voltage ratio, impedance percentage, and voltage regulation.

6-75 If transformers with different electrical characteristics are connected in parallel, what will happen?

They will not distribute the load equally; one transformer will tend to assume more of the load than the other. This leads to overheating, and in severe cases, the destruction of the transformer(s).

CHAPTER 7

Wiring Design and Protection

7-1 What are the minimum requirements for service-drop conductors?

They must be of sufficient size to carry the load that is required of them, but they must not be smaller than No. 8 copper wire or No. 6 aluminum, except under limited load conditions where they may not be smaller than No. 12 hard-drawn copper (see *NEC*, Section 230-23).

7-2 Are the conductors from a pole on which the meter or service switch is installed considered as a service drop?

Yes, and they must be installed accordingly (see *NEC*, Section 230-21).

7-3 What is the minimum clearance for service drops over buildings?

They shall have a minimum clearance of 8 feet [see *NEC*, Section 230-24(a)]. However, if the voltage does not exceed 300 volts between conductors and the roof has a slope of not less than 4 inches in 12 inches, the clearance may be a minimum of 3 feet [see *NEC*, Section 230-24(a), Exception No. 1].

7-4 What is the minimum height of point of attachment of service drops?

10 feet, provided the clearance in the *NEC*, Section 230-24(b), is met (see also *NEC*, Section 230-26).

7-5   What is the minimum clearance of service drops over commercial areas, parking lots, agricultural, and other areas subject to truck traffic?
18 feet [see *NEC*, Section 230-24(b)].

7-6   What is the minimum clearance of service drops over sidewalks?
12 feet [see *NEC*, Section 230-24(b)].

7-7   What is the minimum clearance of service drops over driveways, alleys, and public roads?
18 feet [see *NEC*, Section 230-24(b)].

7-8   What is the minimum clearance of service drops over residential driveways?
12 feet [see *NEC*, Section 230-24(b)].

7-9   Can a bare neutral conductor be buried in the ground in an underground service?
No; it must be insulated, unless it is in a duct or conduit (see *NEC*, Section 230-30). There is an exception: bare copper for direct burial where bare copper is judged to be suitable for the soil conditions.

7-10   Where underground service conductors are carried up a pole, to what minimum height must they be given mechanical protection?
8 feet (see *NEC*, Section 300-5).

7-11   Is sealing required where underground ducts or conduits enter buildings?
Yes, to prevent the entrance of gases or moisture into the building. Spare or unused conduits and ducts must also be sealed (see *NEC*, Section 230-8).

7-12   What is the size of the smallest service-entrance conductor normally allowed?
No. 8 copper or No. 6 aluminum (see *NEC*, Section 230-42, Exceptions).

7-13   What is the minimum size service for a single-family dwelling with an initial load of 10 kVA or more?
The service must be a minimum of 100 amperes for a 3-wire service [see *NEC*, Section 230-42(b)(2)].

7-14   A residence has over six (6) 2-wire branch circuits. What is the minimum size of service required for this residence?
The service entrance conductors shall have an ampacity of not less than 100 amperes for a 3-wire service [see *NEC*, Section 230-42(b)(1)].

7-15   Are splices permitted in service-entrance conductors?
No, they must not be spliced; where they supply a meter, they may be broken as necessary for connecting the meter (see *NEC*, Section 230-46).

7-16   Are conductors other than service-entrance conductors permitted in the same raceways or cables?
No, except for grounding conductors and load management control conductors (see *NEC*, Section 230-7).

7-17 What must be provided for the disconnection of service conductors from building conductors?

Some means must be provided for disconnecting all of the building conductors from the service-entrance conductors, and this means must be located in a readily accessible point nearest the point of entrance of the service conductors, on either the outside or inside of the building, whichever is most convenient [see *NEC*, Section 230-70(a)]. Also see definition in Article 100 for Service Equipment.

7-18 What must the service disconnecting means consist of?

It may consist of not more than six switches or six circuit breakers in a common enclosure or in a group of separate enclosures, provided that they are grouped together. Two or three single-pole switches or circuit breakers may be installed on multiwire circuits and counted as one, provided they have handle ties or handles, so that not more than six operations of the hand are required to disconnect all circuits [see *NEC*, Section 230-71(a)].

7-19 What may be connected ahead of the service-entrance disconnecting switch?

Service fuses, meters, high-impedance shunt circuits (such as potential coils of meters), supply conductors for time switches, surge-protection capacitors, instrument transformers, lightning arresters, solar electric systems, interconnected electric power production sources, and circuits for emergency systems (such as fire-pump equipment, etc.) (see *NEC*, Section 230-82).

7-20 In multiple-occupancy dwellings, is it required that each occupant have access to his disconnecting means?

Yes [see *NEC*, Section 230-72(c)].

7-21 Is it permissible to install overcurrent-protective devices in the grounded service conductor?

Overcurrent-protective devices may not be installed in the grounded service conductor unless a circuit breaker is used that opens all conductors of the circuit simultaneously [see *NEC*, Section 230-90(b)].

7-22 What is the minimum size of service-entrance conductors operating at over 600 volts?

They must not be smaller than No. 6 wire, unless in cable, where they may be a minimum of No. 8 [see *NEC*, Section 230-202(a)].

7-23 What is the maximum circuit protection allowed with flexible cords, sizes No. 16 or No. 18, and tinsel cord?

30 and 20 amperes (see *NEC*, Section 240-4, Exception No. 1).

7-24 What is the maximum overcurrent protection allowed on cords of 10 amperes capacity and 20 amperes capacity?

30-ampere circuit for a 10-ampere cord, and 40-ampere circuits for a 20-ampere cord (see *NEC*, Section 240-4).

7-25 Give the standard ratings (in amperes) for fuses and circuit breakers.

Standard ratings are: 15, 20, 25, 30, 35, 40, 45, 50, 60, 70, 80, 90, 100, 110, 125, 150, 175, 200, 225, 250, 300, 350, 400, 450, 500, 600, 700, 800, 1000, 1200, 1600, 2000, 2500, 3000, 4000, 5000, 6000 (see *NEC*, Section 240-6).

7-26 Where can an overcurrent device be used in a grounded conductor?

No overcurrent device shall be placed in any permanently grounded conductor, except as follows:

Where the overcurrent device simultaneously opens all conductors of the circuit (see Section 240-22 of the *NEC*).

A fuse shall also be inserted in the grounded conductor when the supply is 3-wire, 3-phase AC, one conductor grounded (see Section 430-36 of the *NEC*).

7-27 May fuses be arranged in parallel?

No, except as factory assembled in parallel and approved as a unit (see *NEC*, Section 240-8).

7-28 May breakers be paralleled?

No, only circuit breakers assembled in parallel that are tested and approved as a single unit (see Section 240-8 of the *NEC*).

7-29 Are the secondary conductors of transformer feeder taps required to have overcurrent protection?

Yes; see Exception No. 8 of Section 240-21 of the *NEC*. Also see Section 240-3.

7-30 In what position must a knife switch with fuses be mounted? Why?

In a vertical position, so that when the switch is opened, the blades will not close by gravity; the line side must be connected so that when the switch is opened, the fuses will be deenergized (see Section 240-33 of the *NEC*).

7-31 When overcurrent-protective device enclosures are mounted in a damp location, what precautions must be taken?

The enclosures must be of an identified type for the location and must be mounted with at least a ¼-inch air space between the wall or supporting surface (see *NEC*, Section 240-32).

7-32 Where may Edison-base plug fuses be used?

Plug fuses of the Edison-base type shall be used only for replacement in existing installations where there is no evidence of overfusing or tampering [see *NEC*, Section 240-51(b)].

7-33 What is the maximum voltage rating of plug fuses?

125 volts [see *NEC*, Section 240-50(a)].

7-34 What precautions must be taken if plug fuses are used on new installations?

Fuses, fuseholders, and adapters must be so designed that other Type S fuses may not be used (see *NEC*, Section 240-52).

7-35 What are the size classifications of Type S fuses and adapters?

Not exceeding 125 volts at 0–15 amperes, 16–20 amperes, and 21–30 amperes [see *NEC*, Section 240-53(a)].

7-36 What is a Type S fuse?

A dual element fuse with special threads so that fuses larger than what the circuit was designed for may not be used. They are designed to make tampering or bridging difficult (see *NEC*, Section 240-54).

7-37 What are the overcurrent protection requirements for listed extension cord sets?

See *NEC*, Section 240-4, Exception No. 3.

7-38 Why are systems and circuits grounded?

To limit the excess voltage to ground, which might occur from lightning or exposure to other higher-voltage sources (see *NEC*, Section 250-1).

7-39 Is it necessary to ground one wire on a two-wire DC system with not more than 300 volts between conductors?

Yes, in practically every case, although there are a few exceptions (see *NEC*, Section 250-3).

7-40 When should AC systems be grounded?

AC circuits and systems shall be grounded as provided for in Section 250-5(a), (b), and (c) of the *NEC*.

7-41 When is the next higher overcurrent protective device permitted to protect a conductor where a standard fuse or circuit breaker is not available?

In general up to 800 amperes (see *NEC*, Section 240-3, Exception No. 1 for additional details).

7-42 How are remote control circuits protected against overcurrents?

In accordance with Article 725 (see *NEC*, Section 240-3, Exception No. 4).

7-43 What article in the *NEC* covers protection from overcurrent for industrial machinery?

Article 670 (see *NEC*, Section 240-2).

7-44 When do circuits of 50 volts or less to ground have to be grounded? When do they not have to be grounded?

Circuits of less than 50 volts need not be grounded unless they are supplied from systems exceeding 150 volts to ground, where they are supplied by transformers from ungrounded systems, or where they run overhead outside buildings [see *NEC*, Section 250-5(a)].

7-45 If a grounded conductor is not going to be used in a building, must this grounded conductor be run to each service?

Yes, the grounded conductor must be run to each service. It need not be taken farther than the service equipment if it will not be required in any of the circuits (see *NEC*, Section 250-23).

7-46 Is it necessary to ground each service on a grounded AC system?

Each individual service must have a ground, and this ground must be connected on the supply side of the service-disconnecting means. If there is only one service from a transformer, there must be additional ground connection at the transformer or elsewhere in the transformer circuit for the installation to be approved (see *NEC*, Section 250-23).

7-47 What are the grounding requirements when two or more buildings are supplied from one service?

Where two or more buildings or structures are supplied from a common service, the grounded system in each building or structure is required to have a grounded electrode as described in *NEC*, Section 250-81, connected to the metal enclosure of the building disconnecting means and to the AC system grounded circuit conductor on the supply side of the building or structure disconnecting means (see Section 250-24 of the *NEC* and Exceptions).

7-48 Is it required to ground a separately derived system, and if so, where?

Premises wiring systems that are required by Section 250-5 to be grounded shall be grounded if the phase conductors are not physically connected to another supply system. They shall be grounded at the transformer, generator, or other source of supply, or at the switchboard on the supply side of the disconnecting means (see *NEC*, Section 250-26).

7-49 When must metal, noncurrent-carrying parts of fixed equipment that are likely to become energized be grounded?

When supplied by metal-clad wiring, when located in damp or wet places, when within reach of a person who can make contact with a grounded surface or object or within reach of a person standing on the ground, when in hazardous locations, or where any switches, enclosures, etc., are accessible to unqualified persons (see *NEC*, Section 250-42).

7-50 Is it required to ground metal buildings?

If excessive metal in or on buildings may become energized and is subject to personal contact, adequate bonding and grounding will provide additional safety (see *NEC*, Section 250-44 FPN).

7-51 What is the exception to grounding the noncurrent-carrying parts of portable tools and appliances?

Portable tools and appliances may have double insulation. If they have, they must be so marked and will not require grounding (see *NEC*, Section 250-45).

7-52 What are the grounding requirements when lightning rods and conductors are present?

Metal enclosures with conductors must be kept at least 6 feet away from lightning conductors. Where this is not practical, they must be bonded together (see *NEC*, Section 250-46).

7-53 What is meant by effective grounding?

The path to ground must be permanent and continuous, must be capable of safely handling the currents that may be imposed on the ground, and must have sufficiently low impedance to limit the potential above ground to facilitate the opening of the overcurrent devices. The earth is not permitted to be used as the only equipment grounding conductor (see *NEC*, Section 250-51).

7-54 What is a grounding electrode conductor used for?

The grounding conductor for circuits is used for grounding equipment, conduit, and other metal raceways, including service conduit, cable sheath, etc. (see *NEC*, Section 250-53).

7-55 What appliances may be grounded to the neutral conductor?
Electric ranges and electric clothes dryers, provided that the neutral is not smaller than No. 10 copper wire (see *NEC*, Section 250-60). However, this is not allowed for equipment that is fed from feeder panels, in mobile homes, or in recreational vehicles.

7-56 Three-wire SE cable with a bare neutral is sometimes used for connecting ranges and dryers. Is it permissible to use this type of cable when the branch circuit originates from a feeder panel?
No, the neutral shall be insulated [see *NEC*, Section 250-60(c)].

7-57 What equipment, other than electric ranges and clothes dryers, may be grounded to the grounding conductor?
The grounding conductor on the supply side of the service-disconnecting means may ground the equipment, meter housing, etc. The load side of the disconnecting means cannot be used for grounding any equipment other than electric ranges and dryers (see *NEC*, Section 250-61).

7-58 How should continuity at service equipment be assured?
Threaded couplings and bosses, as well as threadless couplings and connections in rigid conduit or EMT (electrical metallic tubing), must be wrenchtight. Bonding jumpers must be used around concentric and eccentric knockouts and ordinary locknuts and bushings cannot be used for bonding (see *NEC*, Section 250-72).

7-59 On flush-mounted grounded-type receptacles, is it necessary to bond the green grounding screw of the receptacle to the equipment ground?
Yes (see *NEC*, Section 250-74).

7-60 Can conduit serve as the equipment ground?
Except for a few cases, such as special precautions in hazardous locations, conduit, armored cable, metal raceways, etc., can serve as the equipment ground [see *NEC*, Section 250-91(b)].

7-61 What is required to assure electrical continuity of metal raceways and metal-sheathed cable used on voltages exceeding 250 volts?
The electrical continuity of metal raceways or metal-sheathed cable which contains any conductor other than service-entrance conductors of more than 250 volts to ground shall be assured by one of the methods specified in Section 250-72(b), (c), (d), and (e), or by one of the following methods (refer to Section 250-76 of the *NEC*):

(a) Threadless fittings, made up tight, with conduit or metal-clad cable.
(b) Two locknuts, one inside and one outside the boxes and cabinets. Your attention should be specifically called to 250-72(d), which calls for bonding to be used around eccentric or concentric knockouts.

7-62 What is the preferred type of grounding electrode?
A metal underground water system where there are 10 feet or more of buried metal pipe, including well casings that are bonded to the system, is the

preferred grounding electrode [see *NEC*, Section 250-81(a)]. A made electrode shall also be used in addition to the buried metal water pipe.

7-63 Is reinforcing bar in concrete permitted for a grounding electrode when buried metal water piping is not available?
Yes [see *NEC*, Section 250-81(c)].

7-64 What is a "made" electrode?
A made grounding electrode may be a driven pipe, driven rod, buried plate, or other metal underground system approved for the purpose of grounding the equipment (see *NEC*, Section 250-83).

7-65 What are the requirements for plate electrodes?
They must not have less than 2 square feet of surface exposed to the soil. Electrodes of iron or steel must be at least ¼ inch thick; if of nonferrous metal, it must be at least 0.06 inch thick [see *NEC*, Section 250-83(d)].

7-66 What are the requirements for pipe electrodes?
They must be at least ³/₄-inch trade size, and, if made of iron or steel, they must be galvanized or otherwise metal-coated to prevent corrosion; they must also be driven to a depth of at least 8 feet [see *NEC*, Section 250-83(c)].

7-67 What are the requirements for rod electrodes?
Electrodes of iron and steel must be at least ⅝ inch in diameter; if of nonferrous material, they must be listed when ½ inch in diameter. Both types must be driven to a minimum depth of 8 feet [see *NEC*, Section 250-83(c)].

7-68 Describe the installation of made electrodes.
Unless rock bottom is encountered, they must be driven to a minimum depth of 8 feet, and below the permanent-moisture level. Where rock bottom is encountered at a depth of less than 4 feet, they must be buried 2½ feet deep horizontally in trenches or at oblique angles not more than 45 degrees (see *NEC*, Section 250-83).

7-69 What should be the resistance of made electrodes?
If a single made electrode does not measure a resistance to ground of 25 ohms or less, it is required to be supplemented by another electrode which is not less than 6 feet away from the first electrode (see *NEC*, Section 250-84).

7-70 May grounding electrode conductors be spliced?
No, splices are not permitted; grounding electrode conductors must be one piece for their entire length, although there are some exceptions (see *NEC*, Section 250-91).

7-71 What is the smallest size grounding electrode conductor permissible?
No. 4 wire or larger may be attached to buildings, No. 6 wire may be used if properly stapled to prevent physical damage, and No. 8 wire may be used if in conduit or armored cable for protection (see *NEC*, Section 250-92).

7-72 When a grounding electrode conductor is enclosed in a metal enclosure, how is the metal enclosure to be installed?
It shall be electrically continuous from point of attachment to cabinet or enclosure to the point of attachment to the ground clamp or fitting. Metal

enclosures which are not physically and electrically continuous shall be bonded to the grounding electrode conductor at both ends of the metal enclosure (see *NEC*, Section 250-92).

7-73 May aluminum be used for a grounding electrode conductor?

Yes, but it may not come in contact with masonry or the earth and cannot be run closer than 18 inches from the earth [see *NEC*, Sections 250-91(a) and 250-92].

7-74 If a water pipe is used as the grounding electrode, what precautions must be taken?

The grounding connection should be made at the point of water-service entrance. If this cannot be done and there is a water meter on the premises, the water meter is required to be bonded with a jumper of sufficient length, so that the water meter may be readily removed without disturbing the bonding. The cold-water piping is required to be used, and it should be checked to make certain that there are no insulated connections in the piping (see *NEC*, Section 250-112).

7-75 Where nonmetallic water pipe serves the house and a made electrode must be used, what grounding procedures are recommended?

It is required that the ground also be bonded to the interior metal water piping system, including the hot-water piping, and also to the sewer, gas piping, air ducts, etc. This will provide additional safety (see *NEC*, Section 250-80).

7-76 When reinforcing bar is used as the grounding electrode, are clamps required to be listed for connecting the grounding conductor to the rebar?

Yes. Exothermic welding is also permitted (see *NEC*, Section 250-113).

7-77 Can solder be used to attach connections to the grounding conductor?

Solder is never permitted; exothermic welding or listed pressure connectors are to be used (see *NEC*, Section 250-113).

7-78 Flexible metal conduit is generally not permitted for equipment grounding purposes. Is there any exception to this?

Yes, if not over 6 feet long, and the circuit is protected by overcurrent devices of not over 20 amperes (see *NEC*, Section 350-5).

7-79 What means must be taken to maintain continuity at metal boxes when nonmetallic systems of wiring are used, so that the equipment ground wire will be continuous?

The equipment grounds must be attached by means of a grounding screw (boxes are now available with a tapped hole having 10/32 threads) or by some other listed means (see *NEC*, Section 250-114).

7-80 Are sheet-metal straps considered adequate for grounding?

They are not considered adequate unless attached to a rigid metal base that is seated on the water pipe or other ground electrode and listed for the purpose [see *NEC*, Section 250-115(c)].

7-81 What special precautions must be taken in the use of ground clamps?

They must be ot a material suitable for use in connection with the materials that they are attaching to because electrolysis may occur if they are made of different metals. All surfaces must be clean and free of paint or corrosion (see *NEC*, Section 250-115).

7-82 May aluminum grounding conductors be run to cold water piping?

Yes, but any clamps used with aluminum grounding conductors must be approved for the purpose (see *NEC*, Section 250-115).

7-83 May conductors of different systems occupy the same enclosure?

Conductors for light and power systems of 600 volts or less may occupy the same enclosure. The individual circuits may be AC or DC, but the conductors must all be insulated for the maximum voltage of any conductor in the enclosure, except emergency systems [see *NEC*, Section 300-3(c)(1)].

7-84 May conductors of systems over 600 volts occupy the same enclosure as conductors carrying less than 600 volts?

Under certain conditions [see *NEC*, Section 300-3(b)(2), Exceptions No. 1-3].

7-85 If the secondary voltage on electric-discharge lamps is 1,000 volts or less, may its wiring occupy the same fixture enclosure as the branch-circuit conductors?

Yes, if insulated for the secondary voltage involved [see *NEC*, Section 300-3(b),(2), Exception No. 1].

7-86 Is it permissible to run control, relay, or ammeter conductors which are used in connection with a motor or starter in the same enclosure as the motor-circuit conductors?

Yes, if the insulation of all the conductors is enough for the highest voltage encountered [see *NEC*, Section 300-3(b)(2), Exception No. 3].

7-87 When boxes, fittings, conduit, etc., are used in damp or corrosive places, how must they be protected?

They must be protected by a coating of approved corrosion-resistant material [see *NEC*, Section 300-6(b)].

7-88 In damp locations, what precautions must be taken against the corrosion of boxes, fitting, conduit, etc.?

There must be an air space of at least ¼ inch between the wall or supporting material and any surface-mounted conduit, fittings, etc. [see *NEC*, Section 300-6(c)].

7-89 When raceways extend from an area of one temperature into an area of a widely different temperature, what precautions must be taken?

Precautions must be taken to prevent the circulation of air from a warmer to a colder section through the raceway [see *NEC*, Section 300-7(a)].

7-90 When raceways and conductors are run through studs, joints, and/or rafters, what precautions must be taken?

They should be run near the appropriate center of the wood members, or at least 1¼ inches from the nearest edge. If the members have to be notched or a 1¼-inch protection cannot be given, the conduit or conductors must be covered by a steel plate not less than ⅟₁₆ inch in thickness (see *NEC*, Section 300-4).

7-91 On 240/120-volt multiwire branch circuits feeding through an outlet box, may the neutral be broken and connected by means of the screws on the receptacle?

No. In multiwire circuits, the continuity of a grounded conductor shall not be dependent upon a connection device, such as lampholders, receptacles, etc., where the removal of such devices would interrupt the continuity [see *NEC*, Section 300-13(b)].

7-92 How much wire must be allowed at outlets and switch boxes for connections and splices?

There must be at least 6 inches of free conductor left for making the connections (see *NEC*, Section 300-14).

7-93 What precautions must be taken to prevent induced currents in metal enclosures?

When conductors carry AC, all phase wires and the neutral wire, if one is used, and all equipment grounding conductors must run in the same raceway. When single conductors must be passed through metal having magnetic properties (iron and steel), slotting the metal between the holes will help keep down the inductive effect (see *NEC*, Section 300-20).

7-94 When it is necessary to run wires through air-handling ducts or plenums, what precautions must be taken?

The conductors must be run in conduit, electrical metallic tubing, intermediate metal conduit, flexible steel conduit with lead-covered conductors, metal-clad cable, Type MI cable, or intermediate metal conduit (see *NEC*, Section 300-22).

7-95 Does temporary wiring, such as for construction, Christmas lighting, carnivals, etc., come under the *NEC*?

Yes, see *NEC*, Article 305.

7-96 Are ground-fault circuit interrupters required on temporary wiring for construction sites?

Yes, on all 15- and 20-ampere branch circuits [see *NEC*, Section 305-6(a)].

7-97 When conductors are used underground in concrete slabs or other masonry that comes in direct contact with the earth, or where condensation or accumulated moisture in raceways is apt to occur, what characteristic must the insulation have?

It must be suitable for use in a wet location (see *NEC*, Section 310-8).

7-98 Name some insulations permitted for use in a wet location.

RHW, TW, THW, THWN, and XHHW type cable (see *NEC*, Section 310-8).

7-99 What are the cable requirements for buried conductors?

Cables of one or more conductors for direct burial in the earth must be

Type USE cable, except for branch circuits and feeders, which may use Type UF cable. Type UF cable cannot be used for service wires (see *NEC*, Articles 338 and 339).

7-100 What other rules apply to buried conductors?
All conductors, including the neutral, must be buried in the same trench and be continuous (without any splice). Extra mechanical protection may be required, such as a covering board, concrete pad, or raceway (see *NEC*, Section 300-5).

7-101 What is the minimum size of conductors allowed by the *NEC*?
No. 14 wire; however, there are exceptions where smaller sizes may be used (see *NEC*, Section 310-5).

7-102 What is the ruling on stranded conductors?
Except for bus bars, Type MI cable, and pool bonding, conductors of No. 8 and larger must be stranded (see *NEC*, Section 310-3).

7-103 What is the minimum size of conductors that may be paralleled?
Conductors in sizes 1/0 and larger may be run in multiple (see *NEC*, Section 310-4).

7-104 When conductors are run in parallel, what factors must be considered?
Each phase or neutral, if used, must be the same length and of the same conductor material; have the same circular-mil area and the same type of insulation; and be arranged to terminate at both ends so that there will be equal distribution of current between the conductors. Phases must be coordinated to eliminate any induction currents that may be caused to flow in the raceways (see *NEC*, Section 310-4).

7-105 How many conductors may be run in raceways or cables without having to apply a derating factor to the current-carrying capacity?
Three current-carrying conductors (neutrals not included); when more than three current-carrying conductors are run, a derating factor must be applied (see *NEC*, Notes to Tables 310-16 through 310-19).

7-106 What are the derating factors for more than three current-carrying conductors in raceways or cables?

4 to 6 conductors	80% rating
7 to 9 conductors	70% rating
10 to 24 conductors	70% rating
25 to 42 conductors	60% rating
43 conductors and above	50% rating

(See *NEC*, Note 8 to Tables 310-16 through 310-19.)

7-107 In derating, how is the neutral conductor considered?
Normally, the current in the neutral conductor is only the unbalanced current. Therefore, if the system is well balanced, the neutral is not considered as a current-carrying conductor and would not enter into derating (see *NEC*, Note 10 to Tables 310-16 through 310-19).

7-108 In determining the current in the neutral conductor of a wye system, how is the neutral classed?
In a 4-wire, 3-phase, wye-connected system, a common conductor carries approximately the same current as the other conductors and is therefore not considered as a neutral in determining the derating of current capacity (see *NEC*, Note 10 to Tables 310-16 through 310-19).

7-109 What size must the conductors be for a parallel feeder or service?
At least 1/0 and larger for aluminum, copper-clad aluminum, or copper conductors (see *NEC*, Section 310-4).

7-110 What condition permits the use of parallel conductors in sizes smaller than No. 1/0 AWG?
When run for frequencies of 360 hertz and higher (see *NEC*, Section 310-4, Exception No. 3).

7-111 What are the minimum sizes for conductors when operated at a voltage of 0 through 2000 volts?
No. 14 copper, No. 12 aluminum or copper-clad aluminum (see *NEC*, Section 310-5).

7-112 Where installed in raceways, stranded conductors of what size must be used?
No. 8 and larger (see *NEC*, Section 310-3).

7-113 What are the types of conductors required when installed in a wet location?
Lead-covered, types RHW, TW, THW, THWN, XHHW, or types that are listed for wet locations.

7-114 What are the temperature limitations of conductors?
See *NEC*, Section 310-10.

7-115 What are the principal determinants of conductor operating temperatures?
The maximum temperature that the conductor can withstand over a prolonged period of time without serious degradation (see *NEC*, Section 310-10, FPN).

7-116 How are grounded conductors identified?
They must be white or natural gray in color [see *NEC*, Section 310-12(a)].

7-117 How are equipment grounding conductors identified?
They must be green, green with a yellow stripe, or bare [see *NEC*, Section 310-12(b)].

7-118 How are ungrounded conductors identified?
They must be distinguished by colors other than white, gray, or green [see *NEC*, Section 310-12(c)].

7-119 What does the term "electrical ducts" mean?
Suitable electrical raceways, run underground, and embedded in concrete or earth [see *NEC*, Section 310-15(d)].

7-120 What table covers the ampacity of type AC cable?
Table 310-16.

7-121 What is the ampacity of a No. 6 AWG copper type THW conductor?
65 amperes (see *NEC*, Table 310-16).

7-122 What are the ampacity correction factors for a Type AC cable assembly when the ambient temperature is other than 30°C (86°F)?
See *NEC*, Table 310-16.

7-123 What is the maximum operating temperature for a Type THWN conductor?
75°C (167°F) in wet locations (see *NEC*, Table 310-3).

7-124 Under what conditions can a Type XHHW conductor be used in a wet location?
When it is rated at a temperature of 75°C (167°F) (see *NEC*, Section 310-3).

7-125 What are the requirements when more than one calculated or tabulated ampacity could apply for a given circuit length?
The lowest value shall be used [see *NEC*, Section 310-15(c)].

7-126 What part of the *NEC* is used to determine the ampacity of underground runs of conductors?
Table 310-16.

7-127 Give the ampacities for Types AC or NM conductor assemblies based on an ambient air temperature of 30°C (86°F).
See *NEC*, Table 310-16.

7-128 What advantages would two parallel 500-MCM cables have over one 1,000-MCM Cable?
They would be easier to handle and to pull into raceways. According to *NEC*, Table 8, 1,000-MCM Type RH wire has a current-carrying capacity of 545 amperes, whereas 500-MCM cable has a current-carrying capacity of 380 amperes. Therefore, two 500-MCM cables in parallel would have the same circular-mil area as one 1,000-MCM cable and a current-carrying capacity of 760 amperes as against the 545-ampere current-carrying capacity of one 1,000-MCM cable. This is approximately 40% more current-carrying capacity for the two 500-MCM (see *NEC*, Table 8). In addition, there is usually a cost advantage to be gained by using two 500-MCM conductors, rather than one 1,000-MCM conductor.

7-129 May THW insulated conductors be run through a continuous row of fluorescent fixtures?
Yes (see *NEC*, Table 310-13 and Section 410-31).

CHAPTER 8

Wiring Methods and Materials

8-1 Can cable trays be used to support welding cables?
Yes, they must be installed in accordance with *NEC*, Section 630-42 [see *NEC*, Section 318-3(b)(1) FPN].

8-2 Are cable trays intended to be used with ordinary conductors (as described in Article 310 of the *NEC*)?
No, they are not intended for this purpose [see *NEC*, Section 318-3(b)(1)].

8-3 Do cable trays have to be used as a complete system, or may they be used as only partial cable supports?
They must be used as a complete system, including boxes and fittings (see *NEC*, Section 318-6).

8-4 Is grounding required on metallic cable trays?
They must be grounded but must not be used as the neutral conductor. If used as the equipment grounding conductor, they shall meet the requirements of Section 318-7(b) (see *NEC*, Section 318-7).

8-5 Is open wiring on insulators, commonly known as knob-and-tube wiring, recognized by the *National Electrical Code*?
Yes, although you may find that local inspection authorities and regulations may not approve it (see *NEC*, Section 324-3).

8-6   What is Type MI cable?

Type MI cable is a cable in which one or more electrical conductors are insulated with a highly compressed refractory mineral insulation (magnesium oxide) and enclosed in a liquid-tight and gas-tight metallic sheathing (copper) (see *NEC*, Section 330-1).

8-7   Where can Type MI cable be used?

This is one wiring material that can be used for practically every conceivable type of service or circuit. When it is exposed to cinder fill or other destructive corrosive conditions, it must be protected by materials suitable for these conditions (see *NEC*, Section 330-3).

8-8   What precautions must be taken with Type MI cable to prevent the entrance of moisture?

Where Type MI cable terminates, an approved seal must be provided immediately after stripping to prevent the entrance of moisture into the mineral insulation; the conductors must be insulated with an approved insulation where they extend beyond the sheath (see *NEC*, Section 330-15).

8-9   Does the outer sheath of Type MI cable meet the requirements for equipment grounding?

Yes; it provides an excellent path for equipment grounding purposes but not for use as a grounded or neutral conductor [see *NEC*, Sections 330-22, 250-57(a), and 250-91(b)].

8-10   What is armored cable commonly called?

Type AC cable, or in some locations "BX" (see *NEC*, Section 333).

8-11   What is metal-clad cable commonly called?

Type MC cable (see *NEC*, Article 334).

8-12   What is Type MC cable?

Type MC cables are power and control cables in the size range from No. 14 and larger for copper and No. 12 and larger for aluminum and copper-clad aluminum.

The metal enclosures are either a covering of interlocking metal tape or a smooth, impervious, close-fitting, or corrugated tube. Supplemental protection of an outer covering of corrosion-resistant material is required where such protection is needed (see *NEC*, Sections 334-1 and 334-4).

8-13   What is Type AC cable?

Type AC cables are branch-circuit and feeder cables with an armor of flexible metal tape. Cable of the AC type, except Type ACL, must have an internal bonding strip of copper or aluminum in tight physical contact with the armor for its entire length (see *NEC*, Section 333-4).

8-14   At what intervals must Type MC cable be supported?

At intervals not exceeding 6 feet [see *NEC*, Section 334-10(a)].

8-15   At what intervals must Type AC cable be supported?

At intervals of not more than $4\frac{1}{4}$ feet and within 12 inches of boxes and fittings, except where cable is fished and except lengths of not over 2 feet at terminals where flexibility is necessary, and in lengths not over 6 feet for fixture whips (see *NEC*, Section 333-7).

8-16 How may bends be made in Type AC cable?

Bends must be made so that the cables are not damaged. The radius of the curve of the inner edge of any bend must not be less than 5 times the diameter of Type AC cable (see *NEC*, Section 333-8).

8-17 What extra precaution is necessary with Type AC cable when attaching fittings?

An approved insulating bushing or equivalent approved protection must be provided between the conductors and the armor (see *NEC*, Section 333-9).

8-18 What is nonmetallic-sheathed cable?

Nonmetallic-sheathed cable is an assembly of two or more insulated conductors having an outer sheath of moisture-resistant, flame-retardant, nonmetallic material (see *NEC*, Section 336-1).

8-19 What are the available sizes of nonmetallic-sheathed cable?

Sizes No. 14 to No. 2 (AWG) inclusive (see *NEC*, Section 336-25).

8-20 What is the difference between Type NM and Type NMC cables?

In addition to having flame-retardant and moisture-resistant covering, Type NMC cable must also be fungus-resistant and corrosion-resistant [see *NEC*, Section 336-25(b)].

8-21 Where may Type NM cable be used?

It may be used for exposed and concealed work in normally dry locations. It may also be installed or fished in the air voids of masonry block or tile walls, where such walls are not exposed to excessive moisture or dampness [see *NEC*, Section 336-3(a)].

8-22 In dwelling occupancies, what, if any, restrictions have been added?

Types NM and NMC cables shall be permitted to be used in one- and two-family dwellings or multifamily dwellings and other structures not exceeding three floors above grade [see *NEC*, Section 336-4(a)].

8-23 Where may Type NMC cable be installed?

Where exposed to corrosive fumes or vapors, for exposed and concealed work, in dry and moist or damp places, and inside or outside of masonry block or tile walls (see *NEC*, Section 336-3).

8-24 Where is it not permissible to use Type NM or Type NMC cable?

As service-entrance cable, in commercial garages, in theaters, and similar locations, except as provided in the *NEC*, Article 518, "Places of Assembly" [see *NEC*, Section 336-4(a)].

8-25 What supports are necessary for Types NM and NMC cable?

The installation must not be subject to damage, and the cable must be secured in place at intervals not exceeding 4½ feet and supported within 12 inches from boxes and fittings (see *NEC*, Section 336-15).

8-26 What is shielded nonmetallic-sheathed cable?

Type SNM, shielded nonmetallic-sheathed cable, is a factory assembly of two or more insulated conductors in an extruded core of moisture-resistant, flame-retardant, nonmetallic material, covered with an overlapping spiral metal tape and wire shield and jacketed with an extruded moisture, flame, oil, corrosion, fungus and sunlight-resistant nonmetallic material (see *NEC*, Section 337-1).

8-27 Where is SNM cable permitted to be used?
Type SNM cable may be used only as follows:

1. Where operating temperatures do not exceed the rating marked on the cable.
2. In continuous rigid cable trays or in raceways.
3. In hazardous (classified) locations where permitted in Articles 500 through 516 (see *NEC*, Section 337-3).

8-28 What is service-entrance cable?
Service-entrance cable is a conductor assembly provided with a suitable covering, primarily used for services (see *NEC*, Section 338-1).

8-29 What are the two types of service-entrance cable?
Type SE and Type USE (see *NEC*, Section 338-1).

8-30 What is Type SE cable?
It has a flame-retardant, moisture-resistant covering (see *NEC*, Section 338-1).

8-31 What is Type USE cable?
For underground use, it has a moisture-resistant covering but is not required to have a flame-retardant covering (see *NEC*, Section 338-1).

8-32 Can service-entrance cable be used as branch-circuit conductors or feeder conductors?
It may be used for interior wiring systems only when all circuit conductors of the cable are of the thermoplastic or rubber type. However, service-entrance cable without individual insulation on the grounded (neutral) conductor can be used to supply ranges or clothes dryers when the cable does not
• have an outer metallic covering and does not exceed 150 volts to ground. It may also be used as a feeder when it supplies other buildings on the same premises. SE cable with a bare neutral shall not originate in a feeder panel, as covered in Section 250-60(c) (see *NEC*, Section 338-3).

8-33 Describe underground feeder and branch-circuit cable.
Underground feeder and branch-circuit cable must be an approved Type UF cable in sizes No. 14 copper or No. 12 aluminum or copper-clad aluminum to No. 4/0 AWG, inclusive. The ampacity of Type UF cable shall be that of 60°C (140°F) conductors in accordance with Section 310-13. In addition, the cable is permitted to include a bare conductor for equipment grounding purposes only. The overall covering must also be suitable for direct burial in the earth (see *NEC*, Sections 339-1 and 310-15).

8-34 What is required as overcurrent protection for Type UF cable?
Underground feeder and branch-circuit cable may be used underground, including direct burial, provided they have overcurrent protection that does not exceed the capacity of the conductors (see *NEC*, Section 339-3).

8-35 May conductors of the same circuit be run in separate trenches?
All conductors of the same circuit must be run in the same trench or raceway (see *NEC*, Section 339-4).

8-36 What depth must UF cable be buried?
A minimum of 24 inches [see *NEC*, Section 339-3(a)(3) and Table 300-5].

8-37 May UF cable be used for interior wiring?
Yes; however, it must be installed in accordance with Article 336 and must not be exposed to direct sunlight unless it is specifically identified for this purpose (see *NEC*, Section 339-3).

8-38 What is the normal taper on a standard conduit thread-cutting die?
¾ inch per foot [see *NEC*, Section 346-7(b)].

8-39 Where can rigid metal conduit be used?
Practically everywhere; however, rigid metal conduit protected solely by enamel cannot be used outdoors. Also, where practicable, dissimilar metals should not come in contact anywhere in the system to avoid the possibility of galvanic action, which could prove destructive to the system (see *NEC*, Section 346-1).

8-40 When subject to cinder fill, what precautions must be taken with rigid metal conduit?
It must be protected by a minimum of 2 inches of noncinder concrete (see *NEC*, Section 346-3, Exception No. 2).

8-41 What is the minimum size of rigid metal conduit?
½-inch trade size, with a few exceptions (see *NEC*, Section 346-5).

8-42 What are the requirements for couplings on rigid metal conduit?
Threaded or threadless couplings and connectors must be made tight. Where installed in wet locations or buried in concrete masonry or fill, the type used must prevent water from entering the conduit [see *NEC*, Section 346-9(a)].

8-43 Are running threads permitted on rigid metal conduit?
They cannot be used on conduit for connections at couplings [see *NEC*, Section 346-9(b)].

8-44 What is the maximum number of bends permitted in rigid metal conduit between outlets?
Not more than 4 quarter bends, or a total of 360°; this includes offsets, etc., at outlets and fittings (see *NEC*, Section 346-11).

8-45 What is the minimum radius of conduit bends?
See Tables 346-10 and 346-10, Exception, in *NEC*, Table 346-10 is for regular field bends. Table 346-10, Exception is for field bends made with a one-shot bender.

8-46 After a cut is made in rigid metal conduit, what precaution must be taken?
All cut ends of conduits must be reamed to remove any rough ends which might damage the wire when it is pulled in [see *NEC*, Section 346-7(a)].

8-47 **What number of conductors are permitted in conduit?**
See Tables 3A, 3B, and 3C in Chapter 9 of the *NEC*.

8-48 **What are the temperature limitations of conductors?**
See *NEC*, Section 310-10.

8-49 **What are the principal determinants of conductor operating temperatures?**
The maximum temperature that the conductor can withstand over a prolonged period of time without serious degradation (see *NEC*, Section 310-10, FPN).

8-50 **How are grounded conductors identified?**
They must be white or natural gray in color [see *NEC*, Section 310-12(a)].

8-51 **How are equipment grounding conductors identified?**
They must be green, green with a yellow stripe, or bare [see *NEC*, Section 310-12(b)].

8-52 **How are ungrounded conductors identified?**
They must be distinguished by colors other than white, gray, or green [see *NEC*, Section 310-12(c)].

8-53 **What does the term "electrical ducts" mean?**
Suitable electrical raceways, run underground, and embedded in concrete or earth [see *NEC*, Section 310-15(d)].

8-54 **What table covers the ampacity of Type AC cable?**
Table 310-16.

8-55 **What is the ampacity of a No. 6 AWG copper Type THW conductor?**
65 amperes (see *NEC*, Table 310-16).

8-56 **What are the ampacity correction factors for a Type AC cable assembly when the ambient temperature is other than 30°C (86°F)?**
See *NEC*, Section 310-16.

8-57 **What is the maximum operating temperature for a Type THWN conductor?**
75°C (167°F) (see *NEC*, Section 310-13).

8-58 **Under what conditions can a Type XHHW conductor be used in a wet location?**
When it is rated at a temperature of 75°C (167°F) (see *NEC*, Section 310-13).

8-59 **What are the requirements when more than one calculated or tabulated ampacity could apply for a given circuit length?**
The lowest value shall be used [see *NEC*, Section 310-15(c)].

8-60 **What tables are used for underground runs of conductors?**
Tables 310-27 and 310-28 [see *NEC*, Section 310-15(a)].

8-61 **What is intermediate metal conduit?**
It is a metal conduit, similar to rigid metal conduit, but slightly lighter weight (see *NEC*, Article 345).

8-62 What is the minimum size for intermediate metal conduit?
½-inch electrical trade size [see *NEC*, Section 345-6(a)].

8-63 What is the maximum size for intermediate metal conduit?
4-inch electrical trade size [see *NEC*, Section 345-6(b)].

8-64 What are the support requirements for runs of intermediate metal conduit?
See *NEC*, Section 345-12.

8-65 What are the marking requirements for intermediate metal conduit?
See *NEC*, Section 345-16(c).

8-66 What precautions should be taken when installing rigid nonmetallic conduit in extreme cold?
They should not be used in areas where subjected to damage, since they can become brittle at low temperatures (see *NEC*, Section 347-1 FPN).

8-67 What are some of the nonmetallic materials that have been recognized as having suitable physical characteristics when properly formed and treated so that they can be used as a raceway?
See *NEC*, Section 347-1 FPN.

8-68 Should rigid nonmetallic conduit be of a suitable type?
Yes (see *NEC*, Section 347-1).

8-69 Is plastic water pipe permitted to be used as conduit?
It may not be used for this purpose [see *NEC*, Section 110-3(b)].

8-70 What uses are permitted for rigid nonmetallic conduit?
For 600 volts or less, except for direct burial where it is not less than 18 inches below grade and where the voltage exceeds 600 volts, in which case it must be encased in 2 inches of concrete; in concrete walls, floors and ceilings; in locations subject to severe corrosive influences; in cinder fill; and in wet locations (see *NEC*, Article 347-2).

8-71 Where is the use of nonmetallic rigid conduit prohibited?
In hazardous locations, for the support of fixtures or other equipment, where subject to physical damage, or where high temperatures are present (see *NEC*, Section 347-3).

8-72 Where must expansion joints be used on rigid nonmetallic conduit?
Where required to compensate for thermal expansion and contraction (see *NEC*, Section 347-9).

8-73 What is the minimum size of rigid nonmetallic conduit?
No conduit smaller than ½-inch electrical trade size may be used (see *NEC*, Section 347-10).

8-74 Are bushings required when rigid nonmetallic conduit is used?
Yes (see *NEC*, Section 347-12).

8-75 What is electrical metallic tubing (EMT)?
A thin-walled metal raceway (see *NEC*, Article 348).

8-76 May EMT be threaded?
No; EMT can be coupled only by suitable fittings. Where integral couplings are utilized, such couplings shall be permitted to be factory threaded (see *NEC*, Section 348-7).

8-77 In what minimum and maximum sizes is EMT available?
The minimum size is ½ inch with exceptions, and the maximum size is 4 inches, electrical trade size (see *NEC*, Section 348-5).

8-78 May more or fewer conductors be contained in EMT than in rigid metal conduit?
The same number of wires applies to both raceways (see *NEC*, Section 346-6).

8-79 How do the number of bends and the reaming requirements for EMT compare to rigid metal conduit?
They are the same for both (see *NEC*, Sections 348-10 and 348-11).

8-80 When EMT is exposed to moisture or used outdoors, what type of fittings must be used?
They must be raintight fittings (see *NEC*, Section 348-8).

8-81 When EMT is buried in concrete, what type fittings must be used?
They must be concretetight fittings (see *NEC*, Section 348-8).

8-82 What are some common names for flexible metal conduit?
Flex, Greenfield (see *NEC*, Index).

8-83 What is the minimum size for flexible metal conduit?
The normal minimum size is ½ inch, although there are some exceptions to this (see *NEC*. Section 350-3).

8-84 What is the number of wires permitted in flexible metal conduit?
The requirement for this is the same as for rigid metal conduit; however, when ⅜-inch conduit is permitted, a special conductor table is required (see *NEC*, Section 350-3).

8-85 What are the sizes of liquidtight flexible metal conduit?
The sizes of liquidtight flexible metal conduit shall be electrical trade sizes ½ to 4 inches, inclusive. Exception: ⅜-inch size may be used as permitted in the *NEC*, Section 351-5.

8-86 How often should liquidtight flexible metal conduit be supported?
Where liquidtight flexible metal conduit is installed as a fixed raceway, it shall be secured by approved means at intervals not exceeding 4½ feet and within 12 inches on each side of every outlet box or fitting except where conduit is fished, where flexibility is required, and for fixture whips (see *NEC*, Section 351-8).

8-87 May regular flexible metal conduit fittings be used with liquidtight flexible metal conduit?
No, they shall be fittings identified for use with liquidtight flexible metal conduit (see *NEC*, Section 351-7).

8-88 **What is liquidtight flexible nonmetallic conduit?**
See *NEC*, Section 351-22.

8-89 **Where can liquidtight flexible nonmetallic conduit be used?**
See *NEC*, Section 351-23.

8-90 **Under what conditions can liquidtight flexible nonmetallic conduit be used in lengths exceeding 6 feet?**
See *NEC*, Section 351-23(b)(3), Exception.

8-91 **May liquidtight flexible metal conduit be used as an equipment ground?**
Liquidtight flexible metal conduit may be used for grounding in the 1¼-inch and smaller trade sizes if the length is 6 feet or less and it is terminated in fittings listed for grounding (see *NEC*, Section 351-9, Exception No. 1).

8-92 **What are wireways?**
Wireways are sheet-metal troughs with hinged or removable covers for housing and protecting electric wires and cable and in which conductors are laid in place after the wireway has been installed as a complete system (see *NEC*, Section 362-1).

8-93 **What use of wireways is permitted?**
Wireways may be installed only for exposed work; where wireways are intended for outdoor use, they must be of raintight construction (see *NEC*, Section 362-2).

8-94 **What are the uses for which wireways are prohibited?**
Where subject to severe physical damage or corrosive vapors and in hazardous (classified) locations (see *NEC*, Section 362-2).

8-95 **What is the largest size wire permitted in wireways?**
No conductor larger than that for which the wireway is designed shall be installed in any wireway (see *NEC*, Section 362-4).

8-96 **What is the maximum number of conductors permitted in wireways, and what is the maximum percentage of fill?**
A maximum of 30 current-carrying conductors at any cross section; signal circuits or starter-control wires are not included. The maximum fill cannot exceed 20% of the interior cross-sectional area (see *NEC*, Section 362-5).

8-97 **Are splices permitted in wireways?**
Yes, provided that the conductors with splices do not take up more than 75% of the area of the wireway at any point (see *NEC*, Section 362-6).

8-98 **May the wireways be open at the ends?**
No, the dead ends must be closed (see *NEC*, Section 362-9).

8-99 **Is the equipment grounding conductor used in figuring the number of conductors allowed in a box?**
Yes, but only one conductor is to be counted [see *NEC*, Section 370-6(a)(1)].

8-100 What is the purpose of auxiliary gutters?

They are used to supplement wiring spaces at meter centers, distribution centers, switchboards, and similar points of wiring systems; they may enclose conductors or bus bars but cannot be used to enclose switches, overcurrent devices, or other appliances or apparatus (see *NEC*, Section 374-1).

8-101 What is the maximum number of conductors permitted in auxiliary gutters, and what is the maximum percentage of fill allowed?

A maximum of 30 conductors at any cross section; signal circuits and starter control wires are not included. The maximum fill cannot exceed 20% of the interior cross-sectional area (see *NEC*, Section 374-5).

8-102 What is the current-carrying capacity of bus bars in auxiliary gutters?

The current carried continuously in bare copper bars in auxiliary gutters shall not exceed 1000 amperes per square inch of cross section of the conductor. For aluminum bars the current carried continuously shall not exceed 700 amperes per square inch of cross section of the conductor (see *NEC*, Section 374-6).

8-103 May taps and splices be made in auxiliary gutters?

Yes; however, splices may not occupy more than 75% of the cross-sectional area at any point. All taps must be identified as to the circuit or equipment that they supply (see *NEC*, Section 374-8).

8-104 May round boxes be used as outlet, switch, or junction boxes?

No, they cannot be used where conduits or connectors with locknuts or bushings are used (see *NEC*, Section 370-2).

8-105 Is the number of conductors permitted in outlet, switch, and junction boxes limited to any certain number?

Yes; see *NEC*, Tables 370-6(a) and 370-6(b).

Any box less than 1½ inches deep is considered to be a shallow box.

These tables are quite plain; however, they are for conduit or where connectors are used. If cable clamps are used, one conductor must be deducted from the table to allow for the clamp space. Where one or more fixture studs, cable clamps, or hickeys are contained in the box, the number of conductors will be one less than shown in the tables, with a further deduction of one conductor for one or several flush devices mounted on the same strap. A double receptacle is one device; a single switch is one device; three despard switches on one strap is one device; and three despard receptacles on one strap is one device.

There are times when you will have combinations of wire sizes or other conditions, and you will not find the answer in Table 370-6(b). Table 370-6(b) gives the volume of free space (in cubic inches) within a box that is required for each wire size specified. Determine the number of cubic inches in the box; from this you can find the number of various combinations of wire.

8-106 When can nonmetallic boxes be used?

See *NEC*, Section 370-3.

8-107 How are metal boxes grounded?

See *NEC*, Section 370-4.

8-108 What is the size for a square box containing No. 18 or No. 16 conductors?
See *NEC*, Table 370-6(a).

8-109 What is the cubic inch capacity required for a No. 18 conductor?
1.5 cubic inch [see *NEC*, Table 370-6(b)].

8-110 What are the support requirements for boxes and fittings?
See *NEC*, Section 370-13.

8-111 When are exposed surface extensions permitted?
See *NEC*, Section 370-12.

8-112 What are the requirements for pendant boxes?
See *NEC*, Section 370-13(h).

8-113 Can a box be used to support a ceiling fan?
See *NEC*, Sections 370-17(c) and 422-18.

8-114 What protection must be given to conductors entering boxes or fittings?
They must be protected from abrasion, and the openings through which conductors enter must be adequately closed (see *NEC*, Section 370-7).

8-115 May unused openings in boxes and fittings be left open?
No, they must be adequately closed with protection equivalent to that of the wall of the box or fitting (see *NEC*, Section 370-8).

8-116 How far may boxes be set back in walls?
In walls and ceilings of concrete or other noncombustible materials, the boxes may be set back not more than ¼ inch. In walls or ceilings of wood or other combustible materials, the boxes must be flush with the finished surface. If the plaster is broken or incomplete, it must be repaired, so that there will be no gaps or openings greater than ⅛ inch around the box (see *NEC*, Sections 370-10 and 370-11).

8-117 How can you determine dimensions for pull or junction boxes?
For raceways of ¾-inch trade size and larger containing conductors of No. 4 wire or larger, the minimum dimensions of a pull or junction box installed in the raceway for straight pulls must not be less than eight times the trade diameter of the largest raceway. For angle or U pulls, the distance between each raceway entry inside the box and opposite wall of the box must not be less than six times the trade size of the raceway. The distance must be increased for additional entries by the amount of the sum of the diameters of all other raceway entries on the same wall of the box. The distance between raceway entries enclosing the same conductor cannot be less than six times the electrical trade diameter of the larger raceway (see *NEC*, Section 370-18).

8-118 Must pull, outlet, and junction boxes be accessible?
Yes, they must be accessible without removing any part of the building, sidewalk, or paving (see *NEC*, Section 370-19).

8-119 When installing cabinets and cutout boxes in damp or wet places, what precautions must be taken?
They must be installed so that any accumulated moisture will drain out.

There must also be a ¼-inch air space between the enclosure and the surface on which they are mounted (see *NEC*, Section 373-2).

8-120 When mounting cabinets and cutout boxes, how much setback is allowed?

If mounted in concrete or other noncombustible material, they must not be set back more than ¼ inch. If mounted in wood or other combustible material, they must be mounted flush with the finished surface (see *NEC*, Section 373-3).

8-121 Is there any limitation of the deflection of conductors at terminal connections?

Yes, use Table 373-6(a) of the *NEC*. If conductors are deflected or bent too sharply, damage or breakage of conductors may result [see *NEC*, Section 373-5(b)].

8-122 Can unused openings in cabinets or cutout boxes be left open?

No, they must be effectively closed (see *NEC*, Section 373-4).

8-123 What type of bushings must be used on conduits with No. 4 wire or larger?

The bushings must be of the insulating type, or other approved methods must be used when insulating the bushings. Where bonding bushings are also required, combination bonding and insulating bushings are available (see *NEC*, Section 373-6).

8-124 Are meter sockets covered by the *NEC*?

Yes (see *NEC*, Section 373-1).

8-125 What must be done where conductors enter a cabinet or cutout box and are within a cable assembly?

Each cable must be secured to the cabinet or cutout box [see *NEC*, Section 373-5(c)].

8-126 What does the term "offset" mean?

See *NEC*, Section 373-6(b) FPN.

8-127 May enclosures for switches or overcurrent devices be used on raceways or junction boxes?

Enclosures for switches or overcurrent devices shall not be used as junction boxes, auxiliary gutters, or raceways for conductors feeding through or tapping off to other switches or overcurrent devices, unless adequate space is provided so that the conductors do not fill the wiring space at any cross section to more than 40% of the cross-sectional area of the space, and so that the conductors, splices, and taps do not fill the wiring space at any cross section to more than 75% of the cross-sectional area of the space (see *NEC*, Section 373-8).

8-128 May switches be connected in the grounded conductor?

No, unless the ungrounded conductor is opened at the same time as the grounded conductor (see *NEC*, Section 380-2).

8-129 Are signal circuits or control conductors for a motor and starter in auxiliary gutters considered as current-carrying conductors?

No (see *NEC*, Section 374-5, Exception No. 2).

8-130 Can the grounded conductor be switched when using three- or four-way switches?

No, the wiring must be arranged so that all the switching is done in the ungrounded conductor. When the wiring between switches and outlets is run in metal raceways, both polarities must be in the same enclosure (see *NEC*, Section 380-2).

8-131 In what position must knife switches be mounted?

Single-throw knife switches must be mounted so that gravity will not tend to close them. Double-throw knife switches may be mounted either vertically or horizontally; however, if mounted vertically, a locking device must be provided to ensure that the blades remain in the open position when so set (see *NEC*, Section 380-6).

8-132 Do switches and circuit breakers have to be accessible and grouped?

Yes, switches and circuit breakers, as far as is practicable, must be readily accessible (see *NEC*, Section 380-8).

8-133 How must the blades of knife switches be connected?

Unless they are the double-throw-type switches, they must be connected so that when open, the blades will be "dead" (see *NEC*, Section 380-6).

8-134 May knife switches rated at 250 volts at more than 1200 amperes or 600 volts at more than 600 amperes be used as disconnecting switches?

No, they may be used for isolation switches only. For interrupting currents such as these or larger, a circuit breaker or a switch of special design listed for such purpose shall be used (see *NEC*, Section 380-13).

8-135 May fuses be used in parallel on switches?

No, a fused switch shall not have fuses in parallel unless fuses are factory assembled in parallel and approved as a unit (see *NEC*, Section 240-8).

8-136 What is the maximum number of overcurrent devices permitted on one panelboard?

Not over 42 overcurrent devices of a lighting and appliance branch-circuit panelboard can be installed in any one cabinet or cutout box. A two-pole circuit breaker is considered as two overcurrent devices, and a three-pole circuit breaker is considered as three overcurrent devices in the interpretation of the question (see *NEC*, Section 384-15).

8-137 How would you define a lighting and appliance branch-circuit panelboard?

A branch circuit panelboard that has more than 10% of its overcurrent devices rated 30 amps or less (see *NEC*, Section 384-14).

CHAPTER 9

Batteries
and
Rectification

9-1 What is a cell (as referred to in connection with batteries)?

A cell is a single unit capable of producing a DC voltage by converting chemical energy into electrical energy.

9-2 What is a primary cell?

A primary cell is a cell that produces electric current from an electrochemical reaction but is not capable of being recharged.

9-3 What is a secondary cell?

A cell that is capable of being recharged by passing an electric current through it in the opposite direction from the discharging current.

9-4 What is a battery?

Two or more dry cells or storage cells connected together to serve as a single DC voltage source.

9-5 What are the three most common types of cells?

The lead-acid cell, the alkaline cell, and the ordinary dry cell.

9-6 What type of cell is used as an automobile battery?

The lead-acid cell is the one in common use, although the highly expensive and highly dependable nickel-cadmium battery is used to some extent.

9-7 What is the rated voltage of a standard lead-acid cell?
2 volts.

9-8 How are automobile batteries rated?
In ampere-hours.

9-9 Explain the meaning of ampere-hours.
At full charge, the battery is capable of delivering x number of amperes for y number of hours; e.g., a 100-ampere-hour battery would be capable of delivering 10 amperes for 10 hours, 1 ampere for 100 hours, etc.

9-10 What is the specific gravity of the acid in a fully charged (lead-acid) car battery?
Approximately 1.280 to 1.300.

9-11 What is the specific gravity of a discharged car battery?
Approximately 1.200 to 1.215.

9-12 Is it necessary to replace the acid in a car battery?
Under normal conditions, no.

9-13 When adding water to a car battery, what precautions should be observed?
Use distilled water only, and fill the cells only to their prescribed levels.

9-14 Will a discharged car battery freeze easier than a fully charged battery?
Yes.

9-15 Why is it necessary to occasionally recharge a lead-acid battery even though it is not being used?
A lead-acid battery not in use will gradually lose its charge, and if it is left in an uncharged condition, the material on the plates will flake off and short-circuit the plates, thereby causing a shorted cell or cells.

9-16 How may a lead-acid battery be recharged?
By connecting it to a DC source at slightly higher than battery voltage and passing a high current through it in the opposite direction from the discharging current.

9-17 How can you obtain a DC source of power from an AC power source?
By means of AC power rectification.

9-18 Name some rectifiers in common use.
Vacuum-tube rectifiers and semiconductor-diode rectifiers, such as silicon and germanium.

9-19 How does a vacuum-tube rectifier operate?
A vacuum-tube rectifier converts an alternating current into an unidirectional (direct) current.

9-20 How does a contact, or barrier-layer, rectifier work?
It allows passage of current through the contact surface of two materials much more easily in one direction than in the other direction. The contact,

or boundary, surface between the two materials is called the barrier, or blocking, layer.

9-21 Name some barrier-layer rectifiers.
Copper-oxide cells, selenium rectifier cells, magnesium-copper-sulfide cells, and semiconductor rectifiers, such as germanium and silicon.

9-22 Name two common rectifier circuits.
Half-wave rectifier and full-wave rectifier.

9-23 What is a half-wave rectifier?
A rectifier that only passes one-half of an AC wave.

9-24 What is a full-wave rectifier?
A rectifier that passes both halves of an AC wave.

9-25 Draw two cycles of an alternating current. Show the output waveform from a half-wave rectifier and the output waveform from a full-wave rectifier.
See Fig. 9-1.

Fig. 9-1. Two cycles of an alternating current, with the output waveforms from a half-wave rectifier and a full-wave rectifier.

9-26 Can the output wave from half- and full-wave rectification be smoothed out?
Yes, by the proper use of capacitors in the rectifier circuit.

9-27 Name two mechanical methods of producing direct current from alternating current.
A motor-generator set and a rotary converter.

9-28 What is the voltage of the ordinary dry cell?
1½ volts.

9-29 When more voltage than one cell produces is required, how should the cells be connected?

In series.

9-30 When more current is required than one cell can supply, how should the cells be connected?

In parallel.

9-31 How many cells are there in a 12-volt car battery, and how are they connected?

Six cells connected in series. Each cell supplies 2 volts, and, when six cells are connected in series, they supply the required 12 volts.

CHAPTER 10

Voltage Generation

10-1 What is a thermocouple?

A thermocouple consists of two dissimilar metals connected together at one point. When heat is applied to this point, a voltage will be generated.

10-2 Illustrate a thermocouple.

See Fig. 10-1.

10-3 What are some uses of a thermocouple?

As electric thermometers—the voltmeter connected across the circuit is calibrated in degrees of temperature. As a source of electricity on gas furnaces—for operating the valves on the furnace without an external source of power.

10-4 Our normal source of power comes from generators or alternators. By what principle is this power generated in these devices?

It is generated by conductors cutting magnetic lines of force, or by the lines of force cutting the conductors.

10-5 When conductors cut lines of force, what factors determine the voltage generated?

The strength of the magnetic field, the number of conductors in series, and the speed at which the field is cut.

Fig. 10-1.   A diagrammatic representation of a thermocouple.

10-6   Basically, what must a generator consist of?
A magnetic field—usually an electromagnet energized by direct current, an armature or rotor—coils on an iron frame, and some device for taking the current from the rotor or armature—a commutator or slip rings.

10-7   In a standard DC generator, what type of voltage is generated?
The voltage generated is direct current.

10-8   What is the main difference between an alternator and a DC generator?
The method by which the generated power is taken from the generator. The alternator uses slip rings, and the DC generator has a commutator, which takes the current from the coils in such a manner that it always flows in the same direction.

10-9   Draw a simple alternator.
See Fig. 10-2.

Fig. 10-2.   An alternator.

10-10 What does the waveform taken from an alternator look like?
See Fig. 10-3.

10-11 Draw a simple DC generator with only two commutator segments.
See Fig. 10-4.

Fig. 10-3. The waveform from the output of an alternator.

Fig. 10-4. DC generator with two commutator segments.

10-12 What does the waveform from a commutator with only two segments look like?
See Fig. 10-5.

10-13 When more segments are added to the commutator on a DC generator, what happens to the waveform? Explain and illustrate.
The waveform tends to straighten out into a more nearly pure DC waveform (Fig. 10-6).

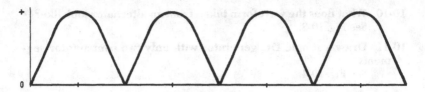

Fig. 10-5. **The waveform from the output of a DC generator having two commutator segments.**

Fig. 10-6. **The addition of commutator segments on a DC generator has the effect of straightening out the output waveform.**

10-14 How is the voltage output on a DC generator regulated?
It may be regulated on changing the strength of the field excitation by means of a rheostat, or by controlling the generator speed.

10-15 How is the voltage output of an alternator controlled?
Since the frequency of the output must be maintained constant, the field excitation must be varied with a rheostat.

10-16 What is the formula for figuring the output frequency of an alternator?

$$\text{Frequency (in hertz)} = \frac{\text{pairs of poles} \times \text{rpm}}{60}$$

10-17 What is the frequency of a 4-pole alternator, driven at 1,800 rpm?
A 4-pole alternator has 2 pairs of poles. Therefore, the frequency is:

$$\frac{2 \times 1,800}{60} = \frac{3,600}{60} = 60 \text{ hertz}$$

10-18 How is the field on a DC generator usually excited?
It is usually self-excited; that is, as the generator starts generating, due to residual magnetism, part of the output is fed into the field.

10-19   How is the field on an alternator usually excited?

A DC generator is usually attached to the end of the alternator shaft. It is from this generator that the excitation current for the alternator field is received.

10-20   On an alternator, does the DC part (the poles) or the AC part (the conductors) rotate?

It is immaterial which part rotates. However, the DC field is usually made the rotating part, and the stator is usually the AC part of the device. This is because the DC field excitation can be of relatively low voltage, and it is easier to insulate a rotating part for low voltage than for high voltage. The AC output is usually a much higher voltage, and it is much more practical to insulate the stator for the high voltage; also, with this arrangement, there are no brushes required on the output side.

10-21   With a rotating DC field on an alternator, what is used for the DC input?

Slip rings are used.

10-22   On a DC generator, which parts, if any, must be laminated? Why?

The armature must be laminated to lower the hysteresis and eddy-current losses. It is not necessary to laminate the field because it is strictly direct current and there would be no hysteresis or eddy-current losses.

10-23   On an alternator, what parts must be laminated? Why?

The stator must be laminated because of hysteresis and eddy-current losses. The rotor need not be laminated because the magnetic flux is steady.

10-24   Why is it preferable to generate alternating rather than direct current?

Alternating current can be changed in voltage by means of transformers; this is necessary because to transmit power over any distance it must be at high voltage due to the losses incurred when transmitting power at low voltage. Direct current cannot be changed in voltage without first changing it to alternating current and then raising the voltage; the operation must then be reversed at the receiving end.

10-25   Is it possible to generate power with a common three-phase, squirrel-cage motor?

Yes; this is what is known as an induction generator.

10-26   How can a common three-phase, squirrel-cage motor be made to generate?

It must be connected in a three-phase circuit and be driven by a prime mover at a faster-than-synchronous speed. It will not generate unless it is connected in a circuit that is already supplying current, or is given a short burst of current to the windings.

10-27   With a squirrel-cage motor connected to line and run above synchronous speed (as described in 10-26), will the motor develop wattless or true power?

It is capable of supplying only true power; the alternator must supply the wattless-power component.

CHAPTER 11

Equipment for General Use

11-1 What may flexible cord be used for?
Flexible cord may be used only for pendants, wiring of fixtures, connection of portable lamps and appliances, elevator cables, wiring of cranes and hoists, for connection of stationary equipment to facilitate their interchange, to prevent the transmission of noise or vibration, and for fixed or stationary appliances where the fastening means and mechanical connections are designed to permit removal for maintenance and repair, or data processing cables as permitted by Section 645-2 of the *NEC* (see *NEC*, Section 400-7).

11-2 What are the prohibited uses of flexible cord?
Flexible cord cannot be used as substitution for the fixed wiring of a structure; run through holes in walls, floors, or ceilings; run through doors, windows, or similar openings; for attaching to building surfaces; or where concealed behind building walls, ceilings, or floors (see *NEC*, Section 400-8).

11-3 Are splices or taps permitted in flexible cord?
No. Flexible cord can be used only in continuous length without splice or tap when initially installed in applications permitted by Section 400-7(a) of the *NEC* (see question 11-1). The repair of hard-service flexible cords No. 14 and larger shall be permitted if conductors are spliced in accordance with Section 110-14(b) of the *NEC*, and the complete splice retains the insulation, outer sheath properties, flexibility, and usage characteristics of the cord being spliced (see *NEC*, Section 400-9).

11-4   What types of flexible cord may be used in show windows and showcases?
Types S, SO, SE, SEO, SJ, SJE, SJO, SJEO, SJOO, ST, STO, STOO, SJT, SJTO, or AFS (see *NEC*, Section 400-11).

11-5   What is the minimum-size conductor permitted for flexible cord?
The individual conductors of a flexible cord or cable shall be no smaller than the sizes shown in Table 400-5 of the *NEC*.

11-6   What are the current-carrying capacities of various flexible cords, in amperes?
Table 400-5 gives the allowable current-carrying capacities for not more than three current-carrying conductors in a cord. If the number of current-carrying conductors in a cord is four to six, the allowable current-carrying capacity of each conductor will be reduced to 80% of the values in the table (see *NEC*, Section 400-5).

11-7   Does the equipment grounding conductor of a flexible cord have to be identified?
Yes, it must be identified with a green color or continuous green color with one or more yellow stripes (see *NEC*, Section 400-23).

11-8   What is the minimum size allowed for fixture wires?
They can be no smaller than No. 18 wire (see *NEC*, Section 402-6).

11-9   What is the ampacity of a copper type W flexible cord sized at No. 8 AWG?
60 amperes (see *NEC*, Section 400-5).

11-10   What are the derating factors to be applied where a flexible cord contains 6 current-carrying conductors?
80 percent of the value given in Table 400-5 (see *NEC*, Section 400-5).

11-11   What does the term "ultimate insulation temperature" mean when applied to flexible cords and cables?
See *NEC*, Section 400-5, Table 400-5 Note.

11-12   What is the ampacity of a 3-conductor type SEO No. 14 flexible cord?
14 amperes (see *NEC*, Section 400-5).

11-13   Is an equipment grounding conductor contained within a flexible cord or cable considered as a current-carrying conductor?
No (see *NEC*, Section 400-5).

11-14   Are flexible cords and cables permitted for temporary wiring on construction sites?
Yes [see *NEC*, Section 400-7(a)(11)].

11-15   What is the allowable current-carrying capacity for fixture wires?
The ampacity of fixture wire shall not exceed the following:

Size (AWG)	Ampacity
18	6
16	8
14	17
12	23
10	28

No conductor shall be used under such conditions that its operating temperature will exceed the temperature specified in *NEC* Table 402-5 for the type of insulation involved.

11-16 What are the requirements, with respect to exposed live parts, on lighting fixtures?

There can be no live parts exposed, except on cleat receptacles and lampholders that are located at least 8 feet above the floor (see *NEC*, Section 410-3).

11-17 What are the requirements for fixtures located in damp areas?

The fixtures must be suitable for the location and must be so installed that moisture cannot enter the raceways, lampholders, or other electrical parts (see *NEC*, Section 410-4).

11-18 What precautions must be taken with fixtures installed near combustible materials?

They must be constructed, installed, or equipped with shades or guards so that combustible materials in the vicinity of the fixture will not be subject to temperatures in excess of 90°C (194°F) (see *NEC*, Section 410-5).

11-19 Can externally wired fixtures be used in show windows?

No, with the exception of chain-supported fixtures (see *NEC*, Section 410-7).

11-20 Where must fixtures be installed in clothes closets?

A fixture in a clothes closet shall be installed:

(1) On the wall above the closet door, provided the clearance between the fixture and the storage area where combustible material may be stored within the closet is not less than 12 inches.

(2) On the ceiling over an area which is unobstructed to the floor, maintaining an 12-inch clearance horizontally between the fixture and a storage area where combustible material may be stored within the closet.

Note: A flush recessed fixture equipped with a solid lens is considered to be outside the closet area.

You are permitted to install a flush recessed fixture that has a solid lens or a ceiling-mounted fluorescent fixture provided it is so mounted or installed that there is a minimum 6-inch clearance horizontally between the fixture and the storage area.

(3) Pendants shall not be installed in clothes closets (see *NEC*, Section 410-8).

11-21 How are flexible cords required to be connected to devices or fittings?

So that no tension will be transmitted to joints or terminals (see *NEC*, Section 400-10).

11-22 Is a conductor within a flexible cord required to be considered as a current-carrying conductor, when it serves as an equipment grounding conductor and a grounding neutral conductor for an electric range or electric clothes dryer?

See *NEC*, Section 400-5.

11-23 What must be done to flexible cords and cables where they pass through holes in covers, outlets boxes, or similar enclosures?

They require protection in the form of bushings (see *NEC*, Section 400-14).

11-24 Under what conditions can a light blue colored insulation be used for the identification of a grounded conductor in a flexible cord?

For jacketed cords furnished with an appliance [see *NEC*, Section 400-22(c)].

11-25 Under what conditions can multiconductor cables rated at over 600 volts, nominal, be used?

To connect mobile equipment and mobile machinery (see *NEC*, Section 402-30).

11-26 What precautions are necessary when installing fixture wires in a colder than 10°C (plus 14°F) temperature?

Exercise care during installation, since the conductors can become brittle (see *NEC*, Section 402-3, FPN).

11-27 What is the ampacity of a No. 10 AWG fixture wire?

28 amperes (see *NEC*, Section 402-5).

11-28 How many fixture wires are permitted in a single conduit?

Not to exceed capacity shown in Table 2 of Chapter 9 (see *NEC*, Section 402-7).

11-29 What is the maximum operating temperature of a Type TFFN fixture wire?

90°C (194°F) (see *NEC*, Table 402-3).

11-30 How are thermoplastic-insulated fixture wires marked?

On their surfaces at intervals not exceeding 24 inches [see *NEC*, Section 402-9(b)].

11-31 Can fixture wire be used as a branch circuit conductor?

No (see *NEC*, Section 402-11).

11-32 What type of fixture is required for a corrosive location?

One that is suitable for such a location [see *NEC*, Section 410-4(b)].

11-33 Are fixtures permitted to be installed in a nonresidential cooking hood?

Yes [see *NEC*, Section 410-4(c)].

11-34 What precautions must be considered when installing a pendant fixture above a bathtub?

It must be at least 8 feet from the top of the bathtub rim and 3 feet beyond the horizontal edge [see *NEC*, Section 410-4(d)].

11-35 What are the requirements for fixtures installed near combustible materials?

They are to be constructed, or installed, or equipped with shades or guards so that combustible material will not be subjected to temperatures in excess of 90°C (194°F) (see *NEC*, Section 410-5).

11-36 Under what condition can a fixture be installed in a show window?

When it is chain-supported and externally wired (see *NEC*, Section 410-6, Exception).

11-37 Does the *NEC* contain a rule that is applicable to space for core lighting?

Yes (see *NEC*, Section 410-9).

11-38 How many requirements must be considered when installing metal poles used to support lighting fixtures?

4 [see *NEC*, Section 410-15(b)].

11-39 What are hickeys, tripods, and crowfeet?

Fittings used to support fixtures [see *NEC*, Section 410-16(d)].

11-40 How are fixtures permitted to be connected to busways?

In accordance with Section 364-12 [see *NEC*, Section 410-16(g)].

11-41 How are fixtures grounded?

By use of an equipment grounding conductor (see *NEC*, Section 410-21).

11-42 Are fixtures required to be provided with a means for the connection of an equipment grounding conductor?

Yes, when they have exposed metal parts (see *NEC*, Section 410-20).

11-43 What are the maximum operating temperatures and voltage limitations for fixture wires?

They are required to be in accordance with Section 402-3 [see *NEC*, Section 410-24(b) FPN].

11-44 Fixtures provided with mogul-base, screwshell lampholders and operating at not over 300 volts between conductors are required to be wired with what type of conductors?

Type AF, SF-1, SF-2, SFF-2, PF, PGF, PFF, PGFF, PTF, PTFF, PAF, PAFF, XF, XFF, ZF, or ZFF fixture wire [see *NEC*, Section 410-25(a)].

11-45 How are fixture conductors and insulation required to be protected?

In a manner that will not tend to cut or abrade the insulation (see *NEC*, Section 410-28).

11-46 Are flexible cords permitted for use to connect fixed showcases in a department store?

Not permitted by *NEC*, Section 410-29.

11-47 When can a fixture be used as a raceway for circuit conductors?

When listed for use as a raceway (see *NEC*, Section 410-31, Exception No. 1).

11-48 Can electric discharge lighting fixtures be connected by a cord and cap?

Yes. When the fixture is located directly below the outlet box, continuously visible, and not subject to strain [see *NEC*, Section 410-30(c)].

11-49 What must be tested before being connected to the circuit supplying a fixture?

All wiring is required to be free from short circuits and grounds (see *NEC*, Section 410-45).

11-50 What are the requirements for the use of double-pole switched lampholders where supplied by ungrounded circuit conductors?

The switching device shall simultaneously disconnect both conductors of the circuit (see *NEC*, Section 410-48).

11-51 Where receptacles rated at 20 amperes or less are directly connected to aluminum conductors what marking is required?

Co/ALR [see *NEC*, Section 410-56(b)].

11-52 A receptacle outlet installed outdoors is required to be located to avoid what condition?

So that water accumulation is not likely to touch the outlet cover or plate [see *NEC*, Section 410-57(e)].

11-53 When installed in a damp or wet location a rosette must be of what type?

Weatherproof type (see *NEC*, Section 410-60).

11-54 What type of wire must be used in the connection of fixtures?

The wire must have the type of insulation that will withstand the temperatures to which the fixture will be exposed. Check the temperature rating of the wire against the temperature at which the fixture operates to obtain the type of insulation required (see *NEC*, Section 410-11).

11-55 May branch-circuit wiring pass through an outlet box that is an integral part of an incandescent fixture?

No, unless the fixture is identified for the purpose (see *NEC*, Section 410-11).

11-56 When a fixture weighs more than 50 pounds, may it be supported from the outlet box?

It must be supported independently of the outlet box [see *NEC*, Section 410-16(a)].

11-57 What is the minimum-size conductor allowed for fixture wires?

No. 18 wire [see *NEC*, Section 410-24(b)].

11-58 What factors must be considered when choosing fixture wires?
Operating temperature, corrosion and moisture conditions, voltage, and current-carrying capacity [see *NEC*, Section 410-24(a)].

11-59 What precautions must be taken in conductors for movable parts of fixtures?
Stranded conductors must be used and must be so arranged that the weight of the fixture or movable parts will not put tension on the conductors. These measures must be taken to protect the conductors (see *NEC*, Section 410-28).

11-60 What types of insulation are permitted within 3 inches of a ballast within a ballast compartment?
90°C (194°F) conductors of the following types: RHH, THHN, FEP, FEPB, SA, XHHW, AVA and THW. See Table 310-13 of the *NEC* for type THW (see *NEC*, Section 410-31).

11-61 Which wire must be connected to the screw shells of lampholders?
The grounded, or white, conductor (see *NEC*, Section 410-23).

11-62 What is the maximum wattage permitted on a medium lamp base?
300 watts (see *NEC*, Section 410-53).

11-63 What is the maximum lamp wattage permitted for a mogul lamp base?
1,500 watts (see *NEC*, Section 410-53).

11-64 For flush or recessed fixtures, what is the maximum operating temperature permitted when installed in or near combustible materials?
90°C (194°F) (see *NEC*, Section 410-65).

11-65 What precautions must be taken on switching of discharge-lighting systems that are rated at 1,000 volts or more?
The switch must be located in sight of the fixtures or lamps, or it may be located elsewhere, provided it is capable of being locked in the open position [see *NEC*, Section 410-81(b)].

11-66 When fixtures are mounted in suspended ceilings, what is required?
Framing members of the suspended ceilings shall be securely fastened together, and the fixtures shall be securely fastened to the ceiling frames by bolting, screws, rivets, or by clips identified for the purpose [see *NEC*, Section 410-16(c)].

11-67 How must fixtures be grounded?
They may be connected to metal raceways, the armor of metal-clad cable, etc., if properly installed and grounded, or a separate equipment grounding conductor not smaller than No. 14 wire may be used (see *NEC*, Section 410-21).

11-68 How must immersion-type portable heaters be constructed?
They must be so constructed and installed that current-carrying parts are effectively insulated from the substance in which they are immersed (see *NEC*, Section 422-9).

11-69 Must electric flatirons have temperature-limiting devices?
Yes (see *NEC*, Section 422-13).

11-70 What type of disconnection means must be provided on stationary appliances?
On appliances rated at less than 300 volt-amperes or ⅛ hp, the branch-circuit overcurrent-protective device may serve as the disconnecting means. For stationary appliances of greater rating, the branch-circuit switch or circuit breaker may serve as the disconnecting means, provided that it is readily accessible to the user of the appliance. For ranges, dryers, and any other cord-connected appliance, the plug at the receptacle may suffice as the disconnecting means (see *NEC*, Section 422-21 and 422-22).

11-71 On motor-driven appliances, what disconnecting means must be provided, and how must it be located?
The switch or circuit breaker that serves as the disconnecting means on a stationary motor-driven appliance must be located within sight of the motor controller or be capable of being locked in the open position (see *NEC*, Section 422-26).

11-72 On all space-heating systems, what is the main requirement?
All heating equipment shall be installed in an approved manner (see *NEC*, Section 424-9).

11-73 What are the requirements for heating cables?
Heating cables must be supplied complete with factory-assembled non-heating leads of at least 7 feet in length (see *NEC*, Section 424-34).

11-74 What markings must be present on heating cables?
Each unit length must have a permanent marking located within 3 inches of the terminal end of the nonheating leads, with the manufacturer's name or identification symbol, catalog number, and the rating in volts and watts or amperes. The leads on a 240-volt cable must be red; on a 120-volt cable the leads are yellow; 208-volt cable leads are blue; and on a 277-volt cable the leads are brown (see *NEC*, Section 424-35).

11-75 With fixed electric space-heating, must controllers or disconnecting means open all ungrounded conductors?
Devices that have an "off" position must open all ungrounded conductors; thermostats without an "on" or "off" position do not have to open all ungrounded conductors or remote control circuits (see *NEC*, Section 424-20).

11-76 What clearance must be given to wiring in ceilings?
Wiring above heated ceilings and contained within thermal insulation must be spaced not less than 2 inches above the heated ceiling and will be considered as operating at 50°C. Wiring above heated ceilings and located

over thermal insulation having a minimum thickness of 2 inches requires no correction for temperature. Wiring located within a joist space having no thermal insulation must be spaced not less than 2 inches above the heated ceiling and will be considered as operating at 50° (see *NEC*, Section 424-36).

11-77 What clearance is required for wiring in walls where electrical heating is used?
Wires on exterior walls must comply with Article 300 and Section 310-10 (see *NEC*, Section 424-37).

11-78 What are the restricted areas for heating?
Heating panels must not extend beyond the room or area in which they originate; cables must not be installed in closets, over cabinets that extend to the ceiling, under walls or partitions or over walls or partitions that extend to the ceiling (see *NEC*, Section 424-38).

11-79 If heating is required in closets for humidity control, may it be used?
Yes, low-temperature heating sources may be used (see *NEC*, Section 424-38).

11-80 What clearance must be provided for heating cables and panels from fixtures, boxes, and openings?
Panels and heating cables must be separated by a distance of at least 8 inches from lighting fixtures, outlets, and junction boxes, and 2 inches from ventilating openings and other such openings in room surfaces, or at least a sufficient area must be provided (see *NEC*, Section 424-39).

11-81 May embedded cables be spliced?
Only when absolutely necessary, and then only by approved means; in no case may the length of the cable be altered (see *NEC*, Section 424-40).

11-82 May heating cable be installed in walls?
No [see *NEC*, Section 424-41(a)].

11-83 What is the spacing requirement on heating cable in dry wall and plaster?
Adjacent runs of heating cable not exceeding 2¾ watts per foot must be installed not less than 1½ inches on centers [see *NEC*, Section 424-41(b)].

11-84 How is heating cable installed in dry board and plaster?
Heating cables may be applied only to gypsum board, plaster lath, or other similar fire-resistant materials. On metal lath or other conducting surfaces, a coat of plaster must be applied first. The entire ceiling must have a coating of ½ inch or more of thermally noninsulating sand plaster or other approved material. Cable must be secured at intervals not exceeding 16 inches with tape, staples, or other approved devices; staples cannot be used with metal lath. On dry-board installations, the entire ceiling must be covered with gypsum board not exceeding ½ inch in thickness, and the void between the upper and lower layers of gypsum board must be filled with thermally conducting plaster or other approved material [see *NEC*, Section 424-41(g)].

11-85 Can the excess of the nonheating leads on heating cable be cut off?

No, they must be secured to the under side of the ceiling and embedded in the plaster. The ends have a color coding on them, and these must be visible in the junction box [see *NEC*, Section 424-43(c)].

11-86 How is heating cable installed in concrete?

Panels or heating units shall not exceed 16½ watts per linear foot of cable. Adjacent runs of cable shall be a minimum of 1 inch apart. The cable shall be secured with frames or spreaders (nonmetallic) while the concrete is being applied. A minimum space of 1 inch must be maintained between the cable and other metallic objects. The leads extending from the concrete shall be protected by rigid metal conduit, intermediate metal conduit, rigid non-metallic conduit, or EMT. Bushings shall be used where cable leads emerge from the floor slab (see *NEC*, Section 424-44).

11-87 What tests must be run on cables during and after installation?

They must be tested for continuity and insulation resistance (see *NEC*, Section 424-45).

11-88 What percentage of branch-circuit ratings are allowable for use with air-conditioning units?

When supplying only the air-conditioning unit, not more than 80% of the circuit rating may be used. When lighting or other appliances are also supplied on the same branch circuit, not more than 50% of the circuit rating may be used for the air-conditioning unit (see *NEC*, Section 440-62).

11-89 Can a plug and receptacle be used as the disconnecting means for an air-conditioning unit?

An attachment plug and receptacle may serve as the disconnecting means for a single-phase, room air-conditioning unit, rated 250 volts or less, when:

(1) The manual controls on the air-conditioning units are readily accessible and located within 6 feet of the floor.

(2) An approved manually operated switch is installed in a readily accessible location within sight of the air-conditioning unit (see *NEC*, Section 440-63).

11-90 What is a sealed (hermetic-type) motor compressor?

This is a mechanical compressor and a motor, enclosed in the same housing, with neither an external shaft nor seal; the motor operates in the refrigerant atmosphere (see *NEC*, Section 440-2).

11-91 What does "in sight from" mean as it is applied to motors?

When "in sight from" is used with reference to some equipment in relation to other equipment, the term means that the equipment must be visible and within 50 feet of each other (see *NEC*, Article 100, for definition).

11-92 How many parts does a motor feeder and branch circuit contain?

Up to nine. See *NEC*, Diagram 430-1.

11-93 What is the purpose of a locked-rotor table?

A locked-rotor table indicates the current, when full voltage is applied to a motor, with the rotor held in a locked position (see *NEC*, Section 430-151).

11-94 How must hermetic-type refrigeration compressors be marked?

They must have a nameplate, giving the manufacturer's name, the phase, voltage, frequency, and full-load current in amperes of the motor current when the compressor is delivering the rated output. If the motor has a protective device, the nameplate must be marked "Thermal Protection." For complete details, see Section 440-4(a) of the *NEC*.

11-95 What markings must be provided on motor controllers?

Maker's name or identification, the voltage, and the current or horse-power rating (see *NEC*, Section 430-8).

11-96 Must the controller be the exact size for the motor it is to be used on?

No, this is not necessary; however, it must be at least as large as is necessary for the job it has to do. It may be larger, but the overloads that might be used in conjunction with it should be the proper size for the application (see *NEC*, Section 430-83).

11-97 When using switches, fuses, and other disconnecting means, must the voltage of the switch coincide with that of the motor?

Yes. The voltage rating of the switch cannot be less than 115% of the full load current rating of the motor (see *NEC*, Section 430-110).

11-98 If a switch is used that has a current rating larger than is required, may it be adapted for less current?

Yes, adapters of the approved type are available for reducing the fuse size of the switch; fuses larger than the switch rating, however, cannot be used.

11-99 Can the enclosures for controllers and disconnecting means for motors be used as junction boxes?

Enclosures for controllers and disconnecting means for motors shall not be used as junction boxes, auxiliary gutters, or raceways for conductors feeding through or tapping off to the other apparatus unless designs are employed which provide adequate space for this purpose [see *NEC*, Section 430-10(a)].

11-100 If a branch circuit supplies only one motor, how large must the conductor size of the branch circuit be in comparison to the current rating of the motor?

The branch-circuit conductors should be figured at not less than 125% of the full-load current rating of the motor (see *NEC*, Section 430-22).

11-101 What are the torque requirements for control circuit devices with screw-type pressure terminals where uses with No. 14 AWG or smaller copper conductors?

They are required to be torqued to a minimum of 7 pound inches [see *NEC*, Section 430-9(c)].

11-102 What type of conductor material is required for connections to motor controllers and terminals of motor control circuit devices?

Copper conductors unless identified for use with a different conductor [see *NEC*, Section 430-9(b)].

11-103 What are the conductor rating factors for power resistors that operate for a continuous duty constant voltage DC motor?
110 percent of the ampacity of conductors (see *NEC*, Section 430-29 and Table 430-29).

11-104 What is a motor control circuit?
See definition in *NEC*, Section 430-71.

11-105 What is the maximum rating of an overcurrent device in amperes for a control circuit conductor sized at No. 18 AWG?
7 amperes [see *NEC*, Section 430-72(b) and Table 430-72(b)].

11-106 What size wire must be used to connect the wound-motor secondary of a motor?
For continuous duty, the conductor size must not be less than 125% of the full-load secondary current of the motor (see *NEC*, Section 430-23).

11-107 When conductors supply several motors, how can you determine the current-carrying capacity of the conductors?
Conductors cannot use less than 125% of the full-load current rating of the highest-rated motor in the group and not less than the full-load current rating of the rest of the motors. The total current is the basis for calculating the wire size (see *NEC*, Section 430-24).

11-108 What overload-current protection must motors of more than 1 hp have?
Motors having a temperature rise of not more than 40°C must trip at not more than 125% of full-load current (see *NEC*, Section 430-32).

11-109 How many conductors must be opened by motor overcurrent devices?
A sufficient number of underground conductors must be opened to stop the motor (see *NEC*, Section 430-38).

11-110 What are the number and location of overcurrent-protection units for different motors?
See *NEC*, Table 430-37.

11-111 How many overcurrent-protective devices are required on a three-phase motor?
One in each phase unless protected by other approved means (see *NEC*, Section 430-37, Exception).

11-112 Where the motor and driven machinery are not "in sight from" the controller location, what conditions must be met?
The controller disconnection means must be capable of being locked in the open position, or there must be, within sight of the motor location, a manually operated switch that will disconnect the motor from its source of supply (see *NEC*, Section 430-102).

11-113 How many motors may be served from one controller?
Each motor must have its own controller; however, where the motors are rated at 600 volts or less and drive several parts of one machine, are under

the protection of one overcurrent-protection device, or where a group of motors is located in one room "in sight from" the controller, a single controller may serve a group of motors (see *NEC*, Section 430-87).

11-114 Must the disconnecting means be marked to indicate the position it is in?
Yes, it must be plainly marked to indicate whether it is in the open or closed position (see *NEC*, Section 430-104).

11-115 May the service switch be used as the disconnecting means?
Yes, if it is "in sight from" the controller location, or if it is capable of being locked in the open position (see *NEC*, Section 430-106).

11-116 Must the disconnecting means be readily accessible?
Yes (see *NEC*, Section 430-107).

11-117 What should be the current-carrying capacity of the disconnecting means?
It must have a current-carrying capacity of at least 115% of the full-load current rating of the motor (see *NEC*, Section 430-110).

11-118 A single-phase motor draws 32 amperes under full-load conditions. How large must the fuses and branch-circuit switch be?
They must be rated at 300% of the full-load current, or about 100 amperes each (see *NEC*, Table 430-150).

11-119 In a group of four motors, one motor draws 10 amperes, one draws 45 amperes, and two draw 75 amperes. What size conductors must be used for the feeder circuit?

$$75 \text{ amperes} \times 125\% = 94 \text{ amperes}$$
$$94 + 75 + 45 + 10 = 224 \text{ amperes}$$

The feeder conductors, therefore, must be able to handle 224 amperes (see *NEC*, Section 430-24).

11-120 What is the full-load current for a synchronous type 75 hp motor operating at 3-phase 460 volts?
78 amperes (see *NEC*, Table 430-150).

11-121 What is the locked rotor current in amperes for a 10 hp motor operating at 3-phase 230 volts?
168 amperes (see *NEC*, Table 430-151).

11-122 Motor locked-rotor currents are approximately how many times the full-load current values given in Tables 430-148 and 430-150?
Six times (see *NEC*, Table 430-151, Note).

11-123 What is the full-load current of a 3 hp motor operating on a single-phase 115-volt system?
34 amperes (see *NEC*, Table 430-148).

11-124 What is the maximum rating or setting of a motor branch-circuit short-circuit and ground fault protective device for a single phase motor with no code letter?

The percent of the full-load current of the motor at 300 percent for a nontime delay fuse; 175 percent for a dual element (time delay) fuse; 700 percent for an instantaneous trip circuit breaker; and 250 percent for an inverse time circuit breaker (see *NEC*, Table 430-152).

11-125 What is the maximum percentage of a full-load current permitted for a high reactance squirrel-cage motor rated at not more than 30 amperes with no code letter, using an instantaneous trip circuit breaker?

700 percent of full load current of the motor (see *NEC*, Table 430-152).

11-126 What are code letters and how are they explained?

See *NEC*, Section 430-7(b) and Table 430-7(b) for code letters and their limitations.

11-127 For small motors not covered in *NEC* Tables 430-147, 430-148, 430-149, or 430-150 what is the locked-rotor current?

They are assumed to be six times the full-load current of the motor [see *NEC*, Section 430-110(c)(3)].

11-128 What is the full-load current for a two-phase 4-wire AC motor sized at 60 hp operating at 230 volts?

133 (see *NEC*, Table 430-149).

11-129 What is the full-load current for a 3-phase AC wound-rotor 100 hp motor operating at a voltage of 460 volts?

124 amperes (see *NEC*, Table 430-150).

11-130 What is the full-load current for a 3-phase AC squirrel-cage 50 hp motor operating at 575 volts?

52 (see *NEC*, Table 430-150).

11-131 Must a controller enclosure have a means for the attachment of an equipment grounding conductor?

Yes (see *NEC*, Section 430-144).

11-132 What is the full-load current in amperes for a 10 hp DC motor operating at 120 volts?

76 amperes (see *NEC*, Table 430-147).

11-133 What is full-load current in amperes for a single-phase AC motor sized at 1 hp operating at 115 volts?

16 amperes (see *NEC*, Section 430-148).

11-134 What is a "controller" when it operates to control a motor?

It is any switch or device that is normally used to stop and start a motor by making and breaking the motor circuit current (see *NEC*, Section 430-81).

11-135 What should the horsepower rating of a controller be?

Not less than the horsepower rating of the motor (see *NEC*, Section 430-83).

11-136 What type of motor controller is used when its submersion is prolonged occasionally?

Enclosure type 6P (see *NEC*, Section 430-91 and Table 430-91).

11-137 Can a motor controller serve as an overload device?

Yes, when the number of overload units complies with *NEC*, Section 430-39.

11-138 What are the requirements for fuses used as overload protection for a motor?

A fuse is required to be inserted in each ungrounded conductor (see *NEC*, Section 430-37).

CHAPTER 12

Motors

12-1 What is a DC motor?

This is a motor designed and intended for use with direct current only.

12-2 If a DC motor is connected across an AC source, what will be the results? Why?

Direct-current flow is only obstructed by resistance, while alternating current is obstructed by both resistance and inductive reactance. Therefore, when a DC motor is connected across an AC source, the current on AC will be much less than that of DC. The motor would run, however, but it would not carry the same load as it would on DC. There would be more sparking at the brushes. The armature is made up of laminations, but the field is not. The eddy currents in the field would therefore cause the motor to heat up and eventually burn on AC; this would not happen on DC.

12-3 What is the difference between a DC motor and a DC generator?

Fundamentally there is none. A DC motor will generate electricity if driven by some prime mover, and a DC generator will act as a motor if connected across a DC source.

12-4 What are the three fundamental types of DC motors?

They are the series-wound motors, the shunt-wound motors, and the compound-wound motors.

12-5 What is a series-wound motor? Draw a schematic of one.

In a series-wound motor, the field is wound in series with the armature (Fig. 12-1).

Fig. 12-1. A series-wound motor.

12-6 What is a shunt-wound motor? Draw a schematic of one.

A shunt-wound motor has the field winding in shunt, or parallel, with the armature. The field consists of many turns of small wire since it must be able to handle the line voltage that is connected across it (Fig. 12-2).

Fig. 12-2. A shunt-wound motor.

12-7 What is a compound-wound motor? Draw a schematic of one.

A compound-wound motor is a combination of a series- and a shunt-wound motor and has better speed regulation than either one (Fig. 12-3).

Fig. 12-3. A compound-wound motor.

12-8 If the field on a DC motor were opened, what would happen?
The motor would try to run away with itself, or, in other words, the motor would reach a very high speed and might destroy itself.

12-9 What are the advantages and disadvantages of DC motors as compared to AC motors?
Speed control of DC motors is much easier, making them more versatile for use where a wide range of speeds is required. They may be used for dynamic braking; that is, a motor on an electric train will act as a motor when required, but when going downhill, it can be used as a generator, thereby putting current back into the line. In generating, it requires power, so it acts as a brake. DC motors, however, require more maintenance than most AC motors. In order to use a DC motor, special provisions must be made.

12-10 What is a motor-generator set?
This is a generator driven by a motor, with both devices mounted on a common base. It might be an AC motor driving a DC generator or a DC motor driving an alternator.

12-11 What is a rotary converter?
This is a self-contained unit having two or more armature windings and a common set of field poles. One armature winding receives the direct current and rotates, thus acting like a motor, while the others generate the required voltage, and thus act as generators.

12-12 If a DC motor is rapidly disconnected from the line, what must be provided?
A means of shorting out the motor, either directly or through a resistance, so that the collapsing magnetic field does not induce a high voltage.

12-13 What different types of alternating current are used on motors?
Single-phase, two-phase, three-phase, and six-phase; 25-cycle, 50-cycle, 60-cycle, and other frequencies may also be used.

12-14 May single-phase motors be run on two- or three-phase lines?
Yes, if they are connected to only two phase wires.

12-15 Can a three-phase motor be run on a single-phase line?
Yes, but a phase splitter must be used.

12-16 What is a phase splitter?
This is a device, usually composed of a number of capacitors connected in the motor circuit, that produces, from a single input wave, two or more output waves which differ in phase from each other.

12-17 When using a phase splitter, there will be some strange current readings noted. What will they be?
At no-load conditions, the current on the three motor leads will be unbalanced: one will have a high current, and the other two will have a low current. As the motor is loaded, these currents will begin to balance out, and at full-load conditions they will have equalized.

12-18 Why would you use a phase splitter and a three-phase motor instead of a single-phase motor?

It is possible that at the moment single-phase power is all that is available, but in the future three-phase power is expected. Therefore, if you purchase the three-phase motor and a phase splitter, the wiring will be in, the motor will be at the desired location, and the expenses will be cut down. There is also less maintenance on three-phase motors; this one fact will often influence the use of a phase splitter.

12-19 What causes a motor to turn?

There are two parts to a motor: a stator, or field, and a rotor, or armature. Around one part there exists a magnetic field from the line current, and in the other part there is an induced current which causes a magnetic field of opposite polarity. These magnetic fields repel one another, thereby causing the rotating member to turn.

12-20 What is termed the "front end" of a motor? Draw a sketch.

The end opposite the shaft is termed the "front end" (Fig. 12-4).

Fig. 12-4. The "front end" of a motor.

SHAFT

FRONT END

12-21 What is the standard direction of rotation for a motor?

Counterclockwise—that is, the shaft of the motor appears to be turning counterclockwise when you are looking at the front end of the motor.

12-22 Illustrate the standard and opposite directions of rotation of a motor.

See Fig. 12-5.

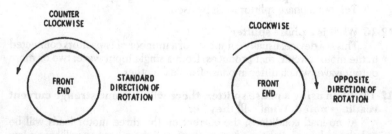

COUNTER CLOCKWISE

CLOCKWISE

FRONT END   STANDARD DIRECTION OF ROTATION

FRONT END   OPPOSITE DIRECTION OF ROTATION

Fig. 12-5. The standard and opposite directions of rotation of a motor.

12-23 What are several types of three-phase motors?
Squirrel-cage, wound-rotor, and synchronous.

12-24 What is a squirrel-cage motor?
This is a motor with the winding on the stator, or line winding. The rotor consists of a winding made of bars that are permanently short-circuited at both ends by a ring.

12-25 How are the voltage and current produced in a squirrel-cage rotor?
The stator can be considered the primary winding, and the rotor the short-circuited secondary winding of a transformer. Thus, the voltage and current are induced in the rotor.

12-26 How is regulation obtained in a motor?
The motor at start is similar to a transformer with a shorted secondary. The current in the rotor and stator will be high. As the motor approaches its rated speed, the rotor induces a voltage into the stator in opposition to the line voltage; this is called *counter emf*. The line current is then reduced in proportion to the speed.

12-27 Does an AC motor (other than a synchronous motor) run at synchronous speed?
No. It must slip below synchronous speed so that there will be an effective voltage produced.

12-28 What is a synchronous motor?
A synchronous motor is almost exactly the same as an alternator. The field must be excited by direct current. The motor runs at the same speed or at a fixed multiple of the speed of the alternator supplying the current for its operation. Should it slip, the motor will pull out and stop since it must run pole for pole with the alternator.

12-29 How do synchronous motors differ from alternators?
They may be just like alternators; however, if they are, they will not be self-starting and will have to be started by some means until they approach synchronous speed, at which time they can be connected to the line and pull into speed. Most synchronous motors have a squirrel-cage winding in addition to the DC field. They start as a squirrel-cage motor; and when they are about up to the speed of the alternator, the DC field is energized. The poles then lock in position with the revolving field of the armature, and the rotor revolves in synchronization with the supply circuit.

12-30 If the field of a synchronous motor is underexcited, what will happen to its power factor?
It will lag.

12-31 If the field of a synchronous motor is overexcited, what will happen to the power factor?
It will lead.

12-32 What is a synchronous capacitor?
It is a synchronous motor running without mechanical load on the line, with its field overexcited for power-factor correction.

12-33 In what direction does the rotor turn in an alternating-current motor?

It always turns in the direction of the rotating field (Fig. 12-6).

Fig. 12-6. The direction of rotor rotation in an AC motor.

12-34 On three-phase motors, how may the fields, or stator windings, be connected internally?

They may be connected in either a delta or a wye arrangement.

12-35 Draw a schematic of a squirrel-cage motor connected in a delta arrangement.

See Fig. 12-7.

Fig. 12-7. A squirrel-cage motor connected in a delta arrangement.

12-36 Draw a diagram of a delta-wound motor. Number the coils and show how they must be connected for a motor that can be used on a 240/480-volt system.

See Fig. 12-8.

Fig. 12-8. A delta-wound motor for use on 240/480 volts.

12-37 How are motor leads numbered?
T_1, T_2, T_3, T_4, etc.

12-38 Draw a diagram of a wye-wound motor. Number the coils and show how they must be connected for a motor that can be used on 240/480 volts.
See Fig. 12-9.

12-39 What is an easy method of remembering which leads to connect together for a wye-wound motor that has two voltage ratings?
First, draw the windings in a wye arrangement. Then, starting at one point, draw a spiral so that it connects in order with all six coils. Begin at the starting point and number from 1 to 9 as you progress around the spiral (Fig. 12-10).

12-40 What are the different kinds of single-phase motors?
Split-phase, capacitor, capacitor-start, capacitor-start-capacitor-run, shading-pole, repulsion, repulsion-induction, and repulsion-start-induction-run.

12-41 What is a wound-rotor, three-phase motor?
This is a three-phase motor that has another three-phase winding instead of a squirrel-cage rotor, the terminals of which are connected to three slip rings. Brushes ride these slip rings and deliver the current to an external three-phase rheostat or variable resistor. At start, all the resistance is in the circuit; as the motor picks up speed, the resistance is gradually decreased until finally the slip rings are short-circuited.

12-42 Can a wound-rotor motor be used as a variable-speed motor?
Yes, if it has been properly designed for this purpose.

Fig. 12-9. A wye-wound motor for use on 240/480 volts.

12-43 How can you reverse the direction of a three-phase squirrel-cage motor?

By transposing any two of the motor leads.

12-44 How can you change the rotation of a wound-rotor motor?

By transposing any two of the line leads.

12-45 Can a wound-rotor motor be reversed by transposing any two leads from the slip rings?

No. There is only one way to reverse its direction and that is by transposing any two line leads.

12-46 Draw a schematic diagram of a wound-rotor motor.

See Fig. 12-11.

12-47 What is a split-phase motor?

This is a single-phase induction motor equipped with an auxiliary winding connected in shunt with the main stator winding and differs from it in both phase and spacing. The auxiliary winding is usually opened by a centrifugal device when the motor has reached a predetermined speed. The fields of these two windings act on the rotor to produce a small starting torque. Once the motor is started, it produces its own rotating field and no longer requires the starting torque.

Fig. 12-10. One easy method of remembering the proper connections for a wye-wound motor with two voltage ratings.

12-48 Does a split-phase motor have a high or low starting torque?
It has a low starting torque.

12-49 How can you reverse the rotation of a split-phase motor?
By reversing the leads to the running or the starting winding, but not both.

12-50 Draw two schematic diagrams illustrating the reversal of a split-phase motor by reversing the running winding leads.
See Fig. 12-12.

Fig. 12-11. The schematic representation of a wound-rotor motor.

Fig. 12-12. The reversal of a split-phase motor is accomplished by reversing the running winding leads.

12-51 What is a capacitor-start motor?

Fundamentally, this motor is very similar to the split-phase motor except that the starting winding has a few more turns and consists of heavier wire than the starting winding of a split-phase motor. There is also a large electrolytic capacitor connected in series with the starting winding. The capacitor and starting winding are cut out of the circuit as soon as the motor reaches approximately 70% of its full speed.

12-52 Does the capacitor-start motor have a high or low starting torque?

It has a high starting torque. For this reason and the fact that it requires little maintenance, it is fast replacing the repulsion types of motors; it is also cheaper to build, and the rewinding is more practical than the repulsion-type motors.

12-53 How can you reverse the direction of rotation of a capacitor-start motor?

By reversing either the running- or starting-coil leads where they are connected to the line. Do not reverse both.

12-54 Are capacitor-start motors usually dual-voltage or single-voltage motors?

They are usually dual-voltage motors.

12-55 If a capacitor-start motor is rated as a 115/230-volt motor (the capacitor is usually a 115-volt capacitor), how is it connected, when the motor is used on 230 volts, to protect the capacitor from being damaged?

The running winding is in two sections so that it may be connected in parallel for 115 volts and connected in series for 230 volts. The capacitor is in series with the starting winding; therefore, when operating at 115 volts, the winding leads are connected across the line at start, and when operating at 230 volts, one lead is connected to one side of the line and the other lead is connected to the midpoint of the running windings.

12-56 Draw two schematic diagrams of a capacitor-start motor, with one showing the 115-volt connection and one showing the 230-volt connection.

See Fig. 12-13.

Fig. 12-13. A capacitor-start motor, first connected across 115 volts and then across 230 volts.

12-57 What is a capacitor-start-capacitor-run motor?

This is the same as the capacitor-start motor except that it has an extra capacitor (oil type) connected in the starting winding, which is always in the motor circuit. The starting capacitor is only in the starting circuit when the motor is started and is disconnected by means of a centrifugal switch.

12-58 What are the advantages of a capacitor-start-capacitor-run motor over the capacitor-start motor?

It is smoother running, has a higher power factor, and consequently draws less current than the capacitor-start motor.

12-59 How can the direction of a capacitor-start-capacitor-run motor be changed?

By reversing the leads to either the starting winding or the running winding, but not both.

12-60 Draw a schematic diagram of a capacitor-start-capacitor-run motor.

See Fig. 12-14.

Fig. 12-14. A capacitor-start-capacitor-run motor.

12-61 Draw a schematic diagram showing the direction reversal of a capacitor-start motor.

See Fig. 12-15.

12-62 Draw a schematic diagram of a two-speed capacitor motor.

See Fig. 12-16.

Fig. 12-15. The reversal of a capacitor-start motor is accomplished by reversing the leads of the running winding.

HI SPEED 3 TO 1, LINE TO 3 & 2
LO SPEED 1 & 3 TO LINE, 2 NO CONNECTION

Fig. 12-16. The schematic representation of a two-speed capacitor motor.

12-63 What is a shaded-pole motor?

This is a single-phase induction motor that is provided with one or more auxiliary short-circuited stator windings that are magnetically displaced from the main winding.

12-64 Does a shaded-pole motor have much starting torque?

No, it has very little starting torque, and for this reason it is used primarily for small fan motors.

12-65 In what direction do shaded-pole motors run?

They rotate toward the shading coil.

12-66 Draw a sketch of one pole of a shaded-pole motor showing the direction of rotation.

See Fig. 12-17.

12-67 What is a repulsion motor?

This motor has a stator like that of most single-phase motors and a rotor similar to that of a DC motor. It has a field and an armature; however, the armature is not connected to the supply source, and the brushes are short-circuited, or connected together, by a conductor of negligible resistance.

12-68 Draw a schematic diagram of a repulsion motor.

See Fig. 12-18.

12-69 What is a repulsion-induction motor?

This is a repulsion motor with a squirrel-cage rotor on the armature, so that after the motor starts, it will run as an induction motor.

LEADS

SHADED COIL USUALLY 1 TURN
OF HEAVY COPPER WIRE
SHORT CIRCUITED

ROTATION

Fig. 12-17. One pole of a shaded-pole motor.

LINE

ARMATURE

Fig. 12-18. The schematic representation of a repulsion motor.

12-70 What is a repulsion-start-induction-run motor?
This is a repulsion motor that has a centrifugal-force device to short-circuit the commutator when the motor reaches its rated speed.

12-71 How can the direction of rotation for the repulsion, repulsion-induction, and repulsion-start-induction-run motors be changed?
By shifting the position of the brushes about 15° (electrical).

12-72 What is a universal motor?
This is a motor built like a series DC motor. However, the stator and armature are both laminated, designed for high speeds, and may be used on either AC or DC, although the speed and power will be greater on DC.

12-73 How can the direction of rotation of a universal motor be changed?
By reversing either the field leads or the armature leads, but not both.

12-74 How can the direction of rotation of a DC motor be changed?
By reversing either the field leads or the armature leads, but not both.

12-75 Knowing the frequency and number of poles of an AC motor, how can its speed be determined? This will be synchronous speed; the

actual speed will be slightly lower.
The speed of an AC motor can be found by using the formula

$$\text{Speed} = \frac{120 \times \text{frequency}}{\text{number of poles}}$$

12-76 We have a 60-cycle, 4-pole AC motor. What will its speed be?

$$\text{Speed} = \frac{120 \times 60}{4} = \frac{7,200}{4} = 1,800 \text{ rpm}$$

CHAPTER 13

Motor Controls

Motor controls are not really complicated. Once the fundamental idea is mastered, all types of controls can be figured out and sketched. First, you must become familiar with the common symbols that are used in connection with controls; second, analyze what you want to accomplish with the particular control. If you follow through, your diagram will fit right into place.

There are two common illustrative methods—*diagrams* and *schematics*. Both are used, and in practically every case either one will be acceptable to the examiner. However, you should ask the examiner to be certain that you may use either; this gives you the opportunity to use the one you are most familiar with. The schematic method, however, presents a clearer picture of what you are trying to accomplish and is easier for the examiner to follow and correct. Whichever method you use take your time and make a clear sketch. There is nothing so irritating to an examiner as trying to determine what your intent was and whether you had any idea of what you were doing.

Common Symbols for Motor Controls

PILOT LAMP

PILOT LAMP WITH PUSHBUTTON TO TEST

NORMALLY OPEN CONTACTOR WITH BLOWOUT

NORMALLY CLOSED CONTACTOR WITH BLOWOUT

NORMALLY OPEN CONTACTOR

NORMALLY CLOSED CONTACTOR

SHUNT COIL

SERIES COIL

THERMAL OVERLOAD RELAY

MAGNETIC RELAY

LIMIT SWITCH, NORMALLY OPEN

LIMIT SWITCH, NORMALLY CLOSED

FOOT SWITCH, NORMALLY OPEN

FOOT SWITCH, NORMALLY CLOSED

VACUUM SWITCH, NORMALLY OPEN

VACUUM SWITCH, NORMALLY CLOSED

LIQUID-LEVEL SWITCH, NORMALLY OPEN

LIQUID-LEVEL SWITCH, NORMALLY CLOSED

TEMPERATURE-ACTUATED SWITCH, NORMALLY OPEN

TEMPERATURE-ACTUATED SWITCH, NORMALLY CLOSED

FLOW SWITCH, NORMALLY OPEN

FLOW SWITCH, NORMALLY CLOSED

MOMENTARY-CONTACT SWITCH, NORMALLY OPEN

MOMENTARY-CONTACT SWITCH, NORMALLY CLOSED

IRON-CORE INDUCTOR

AIR-CORE INDUCTOR

SINGLE-PHASE AC MOTOR

3-PHASE, SQUIRREL-CAGE MOTOR

2-PHASE, 4-WIRE MOTOR

WOUND-ROTOR, 3-PHASE MOTOR

ARMATURE

CROSSED WIRES, NOT CONNECTED

CROSSED WIRES, CONNECTED

FUSE

THERMOCOUPLE

DIODE (rectifier)

CAPACITOR

ADJUSTABLE CAPACITOR

RESISTOR

TAPPED RESISTOR

VARIABLE RESISTOR

RESISTOR WITH TWO TAPS

WIRING TERMINAL

FULL-WAVE RECTIFIER

MECHANICAL INTERLOCK

MECHANICAL CONNECTION

13-1 Draw a *diagram* of a magnetic three-phase starter with one start-stop station.
    See Fig. 13-1.

Fig. 13-1.   A diagrammatic representation of a magnetic three-phase starter with one start-stop station.

13-2 Draw a *schematic* of a magnetic starter with one start-stop station.
    See Fig. 13-2. Note that Figs. 13-1 and 13-2 both accomplish the same function, but you can see why the schematic method is preferred.

13-3 What happens on the normal magnetic starter if the electricity drops off for an instant?
    The magnetic starter will drop out and will stay out until restarted.

13-4 How can you restart the magnetic starter after a momentary lightning interruption?
    The holding contacts (M in Fig. 13-2) are shunted with a timing device that maintains the contact for a certain time period, which is ordinarily adjustable for any desired delay.

13-5 In an area where a number of timing devices are used, what would happen if they were all set for the same time-delay interval?
    They would all come back on at the same instant, thereby causing a tremendous current surge on the distribution system, which would cause the line breakers to trip. This situation can be avoided by setting the time delays at different intervals of time, so that they will be staggered when coming back in the system.

13-6 Draw a maintained-contact control, and explain how it differs from Fig. 13-2.

   The maintained-contact control has no voltage drop-out device; therefore, if the current is interrupted, the starter will come back on the line when electric service is restored and will function normally until another interruption (Fig. 13-3).

Fig. 13-2.   A schematic representation of a magnetic three-phase starter with one start-stop station.

Fig. 13-3.   A maintained-contact control.

13-7 Name a few devices that may be used as the pilot in Fig. 13-3.

   Limit switch, pressure switch, thermostat, flow switch, low-water switch, and hand-operated switch.

13-8 How can you use low voltage to control a motor starter that is designed for use on high voltage?

   A transformer or other source of voltage could be used if the control wires are run in the same raceways as the motor conductors. They must have insulation equal to the proper voltage rating of the highest voltage used.

13-9 Draw a schematic of a low-voltage control using a transformer.

   See Fig. 13-4.

13-10 Draw a schematic of a low-voltage control using a control relay to energize the magnetic coil with full voltage.
See Fig. 13-5.

13-11 Draw a schematic of a magnetic starter with one stop-start station and a pilot that burns when the motor is running.
See Fig. 13-6.

Fig. 13-4. A low-voltage motor control using a transformer.

Fig. 13-5. A low-voltage motor control using a transformer in conjunction with a control relay.

13-12 Draw a schematic of a magnetic starter with three stop-start stations.
See Fig. 13-7.

13-13 What is jogging?
This means inching a motor, or constant starting and stopping of the motor to move it a little at a time.

13-14 Draw a starter schematic with a jogging switch.
See Fig. 13-8.

Fig. 13-6. A magnetic starter with one stop-start station and a pilot lamp which burns to indicate that the motor is running.

Fig. 13-7. A magnetic starter with three stop-start stations.

13-15 What is plugging?
This means stopping a motor by instantaneously reversing it until it stops.

13-16 Draw a starter schematic illustrating connections for plugging a motor provided with a safety latch.
See Fig. 13-9.

13-17 Draw a schematic of three motors that are all started and stopped from one stop-start station so that they will all stop if one overload trips.
See Fig. 13-10.

Fig. 13-8. A magnetic starter with a jogging switch.

Fig. 13-9. A magnetic starter with a plugging switch.

13-18 Draw a schematic of two magnetic starters controlled from one start-stop station, with a starting time delay between the two motors.
See Fig. 13-11.

13-19 Draw a schematic of three separately started motors, all of which may be stopped by one master stop station or stopped if one overload trips.
See Fig. 13-12.

13-20 Draw a schematic of a starting compensator, showing start and run contacts.
See Fig. 13-13.

13-21 Draw a diagram of a 200-ampere service with a 12½-kVA alternator as a standby; this will be a nonautomatic switchover. The diagram

Fig. 13-10. Three motors that are simultaneously controlled by one stop-start station; if one overload device trips, all three motors will stop running.

Fig. 13-11. Two magnetic starters that are controlled by one start-stop station; there is a starting-time delay device between the two motors.

must be arranged so that the two sources cannot be connected together, and both must have proper overload protection.

See Fig. 13-14.

13-22 Draw a schematic of a two-speed, 3-phase, squirrel-case motor starter; show motor connections.

See Fig. 13-15.

Fig. 13-12. Three separately started motors which may be stopped by
one master stop station or stopped if one overload device
is tripped.

Fig. 13-13. A schematic representation of a starting compensator,
with start and run contacts.

13-23 What is a starting compensator?

This is an AC device that consists of a built-in autotransformer, which
reduces the voltage to the motor at start. After coming up to partial speed at
a reduced voltage and reduced line current, the motor is connected across the
line for its running position (see Fig. 13-13 for the magnetic-type compensator).

13-24 Draw a diagram of a manual-type starting compensator.

See Fig. 13-16.

Fig. 13-14. A 200-ampere service with a 12½ -kVA standby alternator arranged for a nonautomatic switchover.

13-25 Draw a diagram of a two-speed AC motor control with pushbutton control.

See Fig. 13-17.

13-26 Draw a diagram of the control arrangement for an AC multispeed motor.

See Fig. 13-18.

13-27 Draw a wiring diagram for a DC magnetic starter with a contactor and overload relay.

See Fig. 13-19.

13-28 Draw a wiring diagram of a DC shunt motor speed-regulating rheostat for starting and speed control by field control.

See Fig. 13-20.

13-29 Draw a wiring diagram of a DC speed-regulating rheostat for shunt or compound-wound motors with contactors and pushbutton station control.

See Fig. 13-21.

Fig. 13-15. A two-speed, three-phase, squirrel-cage motor starter.

13-30 Draw a wiring diagram of a DC speed-regulating rheostat for shunt or compound-wound motors without a contactor.

See Fig. 13-22.

13-31 Draw a diagram of a DC speed-regulating rheostat for shunt or compound-wound motors. Regulating duty—50% speed reduction by armature control and 25% increase by field control.

See Fig. 13-23.

13-32 Draw a diagram of magnetic controller for a constant-speed DC shunt or compound-wound nonreversible motor with dynamic braking.

See Fig. 13-24.

Fig. 13-16. A manual-type starting compensator.

13-33 Draw a magnetically operated DC motor started with three pushbutton control stations.

See Fig. 13-25.

13-34 Draw a pushbutton-operated DC motor starter showing a starting arrangement wherein the armature starting current is limited by a step-by-step resistance regulation.

See Fig. 13-26.

Fig. 13-17. A two-speed AC motor with pushbutton control.

Fig. 13-18. The control arrangement for an AC multispeed motor.

Fig. 13-19. The wiring diagram for a DC magnetic starter, with a contactor and an overload relay.

Fig. 13-20. A DC shunt motor speed-regulating rheostat for starting and speed control by field control.

Fig. 13-21. A DC speed-regulating rheostat control for shunt or compound-wound motors with contactors and pushbutton station control.

Fig. 13-22. A DC speed-regulating rheostat control for shunt or compound-wound motors withut a contactor.

Fig. 13-23. A DC speed-regulating rheostat control for shunt or compound-wound motors; regulating duty—50% speed reduction by armature control and 25% increase by field control.

Fig. 13-24. A magnetic controller for constant-speed DC shunt or compound-wound nonreversible motors with dynamic braking.

Fig. 13-25. A magnetically operated DC motor starter with three push-button control stations.

Fig. 13-26. A pushbutton-operated DC motor starter in which the armature starting current is limited by a step-by-step resistance regulation.

CHAPTER 14

Special Occupancies and Hazardous (Classified) Locations

14-1 Where rigid conduit is used in hazardous (classified) locations, what precautions must be taken?

All threaded connections must be made up wrench tight to minimize sparking when fault currents flow through the conduit system. Where this is not possible, a bonding jumper must be used (see *NEC*, Section 500-2).

14-2 Various atmospheric mixtures require different treatment. What is the grouping given to these various mixtures?

The characteristics of various atmospheric mixtures of hazardous gases, vapors, and dusts depend on the specific hazardous material involved. It is necessary, therefore, that equipment be approved not only for the class of location but also for the specific gas, vapor, or dust that will be present.

For purposes of testing and approval, various atmospheric mixtures have been grouped on the basis of their hazardous characteristics, and facilities have been made available for testing and approval of equipment for use in the following atmospheric groups:

Group A—Atmospheres containing acetylene;

Group B—Atmospheres containing hydrogen, or gases or vapors of equivalent hazard such as manufactured gas;

Group C—Atmospheres containing ethylene, or cyclopropane;

Group D—Atmospheres containing gasoline, hexane, naphtha, benzene, butane, propane, alcohol, acetone, benzol, lacquer solvent vapors, or natural gas;

Group E—Atmospheres containing combustible metal dust;

Group F—Atmospheres containing carbon black, coal, or coke dust;

Group G—Atmospheres containing flour, starch, or grain dust.

See the Fine-Point Notes of Section 500-3 in the NEC for complete coverage.

14-3 How shall equipment for hazardous (classified) locations be marked?

Approved equipment shall be marked to show the class, group, and operating temperature or temperature range, based on operation in a 40°C ambient for which it is approved.

The temperature range, if provided, shall be indicated in identification numbers as shown in NEC Table 500-3(b).

Identification numbers marked on the equipment nameplates shall be in accordance with NEC Table 500-3(b).

Exception: Equipment of the nonheat-producing type, such as junction boxes, conduit, and fittings, are not required to have a marked operating temperature.

14-4 What is a Class I location?

These are locations in which flammable gases or vapors are, or may be, present in the air in quantities sufficient to produce explosive or ignitible mixtures (see NEC, Section 500-5).

14-5 What is a Class I, Division 1 location?

Locations in which hazardous concentrations of flammable gases or vapors exist continuously, intermittently, or periodically under normal operating conditions; in which hazardous concentrations of such gases or vapors may exist frequently because of repair or maintenance operations or because of leakage; or in which breakdown or faulty operations of equipment or processes that might release hazardous concentrations of flammable gases or vapors might also cause simultaneous failure of electrical equipment [see NEC, Section 500-5(a)].

14-6 What is a group A atmosphere in a hazardous (classified) location?

One that contains acetylene (see NEC, Section 500-3, FPN No. 5).

14-7 What is a group C atmosphere in a hazardous (classified) location?

One that contains cyclopropane, ethyl ether, ethylene, or gases or vapors of equivalent hazard (see NEC, Section 500-3, FPN No. 7).

14-8 What is intrinsically safe wiring and equipment?

Equipment and wiring that is not capable of releasing sufficient electrical or thermal energy under normal or abnormal conditions to cause ignition of a specific flammable or combustible atmospheric mixture in its most easily ignitible concentration (see *NEC*, Section 500-2).

14-9 What precautions must be taken when installing intrinsically safe circuits?

They are required to be physically separated from wiring of all other circuits that are not intrinsically safe (see *NEC*, Section 500-2).

14-10 Are locations where pyrophoric materials are the only materials used or handled required to be classified as hazardous?

No (see *NEC*, Section 500-2).

14-11 What is pyrophoric material?

A pyrophoric material is one that is capable of igniting spontaneously when exposed to air.

14-12 What is a group G atmosphere?

One that contains combustible dusts having resistivity of 108 ohm-centimeter or greater.

14-13 What is required when it is impractical to make a threaded joint tight in a run of rigid metal conduit installed in a hazardous (classified) location?

A bonding jumper is required to be installed (see *NEC*, Section 500-2).

14-14 What is the maximum temperature that will not exceed the ignition temperature of a specific gas or vapor that is encountered in a hazardous (classified) location?

See *NEC*, Section 500-3(c) and Table 500-3(b).

14-15 What is a Class I, Division 2 location?

Locations in which flammable volatile liquids or flammable gases are handled, processed, or used, but in which the hazardous liquids, vapors, or gases will normally be confined within closed containers or closed systems from which they can escape only in case of accidental rupture or breakdown of such containers or systems, or in case of abnormal operation of equipment; in which hazardous concentrations of gases or vapors are normally prevented by positive mechanical ventilation, but which might become hazardous through failure or abnormal operation of the ventilating equipment; or which are adjacent to Class I, Division 1 location and to which hazardous concentrations of gases or vapors might occasionally be communicated unless such communication is prevented by adequate positive-pressure ventilation from a source of clean air, and effective safeguards against ventilation failure are provided [see *NEC*, Section 500-5(b)].

14-16 What are Class II locations?

Those which are hazardous because of the presence of combustible dust (see *NEC*, Section 500-6).

14-17 What are Class II, Division 1 locations?

Locations in which combustible dust is, or may be, in suspension in the air continuously, intermittently, or periodically under normal operating conditions, in quantities sufficient to produce explosive or ignitible mixtures; where mechanical failure or abnormal operation of machinery or equipment might cause such mixtures to be produced, and might also provide a source of ignition through simultaneous failure of electrical equipment, operation of protective devices, or from other causes; or in which dusts of an electrical conducting nature may be present [see *NEC*, Section 500-6(a)].

14-18 What are Class II, Division 2 locations?

Locations in which combustible dust will not normally be in suspension in the air or is likely to be thrown into suspension by the normal operation of equipment or apparatus, in quantities sufficient to produce explosive or ignitible mixtures, but where deposits or accumulations of such dust may be sufficient to interfere with the safe dissipation of heat from electrical equipment and apparatus, or where such deposits or accumulations of dust on, in, or in the vicinity of electrical equipment might be ignited by arcs, sparks, or burning material from such equipment [see *NEC*, Section 500-6(b)].

14-19 What are Class III locations?

Locations which are hazardous because of the pressure of easily ignitible fibers or flyings, but in which such fibers or flyings are not likely to be suspended in the air in quantities sufficient to produce ignitible mixtures (see *NEC*, Section 500-7).

14-20 What are Class III, Division 1 locations?

Locations in which easily ignitible fibers or materials producing combustible flyings are handled, manufactured, or used [see *NEC*, Section 500-7(a)].

14-21 What are Class III, Division 2 locations?

Locations in which easily ignitible fibers are stored or handled, except in process or manufacture of the product [see *NEC*, Section 500-7(b)].

14-22 Where may meters, instruments, and relays be mounted in Class I, Division 1 locations?

Within enclosures approved for this location [see *NEC*, Section 501-3(a)].

14-23 Where may meters, instruments, and relays with make-and-break contacts be mounted in Class I, Division 2 locations?

Immersed in oil or hermetically sealed against vapor or gases [see *NEC*, Section 501-3(b)].

14-24 What type of wiring must be used in Class I, Division 1 areas?

Rigid metal conduit (threaded), intermediate metal conduit, or Type MI cable [see *NEC*, Section 501-4(a)].

14-25 What type of fittings must be used in Class I, Division 1 areas?

All boxes, fittings, and joints must be threaded. At least five full threads must be used and must be fully engaged, and all materials, including flexible connections, must be approved (explosionproof) by the inspecting authority for these locations [see *NEC*, Section 501-4(a)].

14-26 What type of wiring must be used in Class I, Division 2 areas?
Threaded rigid metal conduit, threaded steel intermediate metal conduit, enclosed gasketed busways, or Type PLTC cable in accordance with the provisions of Article 725, Type MI, MC, MV, TC, or SNM cable with approved termination fittings [see *NEC*, Section 501-4(b)].

14-27 What type of fittings must be used in Class I, Division 2 areas?
Only boxes, fittings, and joints that are approved for this location [see *NEC*, Section 501-4(b)].

14-28 Why is sealing needed in conduit systems and not in Type MI cable systems in Class I areas?
Seals are provided in conduit systems to minimize the passage of gases, vapors, or fumes from one portion of the electrical system to another through the conduit. Type MI cable is inherently constructed to prevent the passage of gases, etc., but sealing compound is used in cable-termination fittings to exclude moisture and other fluids from the cable insulation (see *NEC*, Section 501-5).

14-29 When connecting conduit to switches, circuit breakers, etc., where must seals be placed?
As close as possible to the enclosure, but not more than 18 inches away (see *NEC*, Section 501-5).

14-30 When a conduit leaves a Class I, Division 1 location, where must the seal be located?
The seal must be at the first fitting when a conduit leaves this area; it may be on either side of the boundary [see *NEC*, Section 501-5(a)].

14-31 When a conduit leaves a Class I, Division 2 location, where must the seal be located?
The seal must be at the first fitting when a conduit leaves this area; it may be on either side of the boundary [see *NEC*, Section 501-5(b)].

14-32 If there is a chance that liquid will accumulate at a seal, what precautions must be taken?
An approved seal for periodic draining of any accumulation must be provided [see *NEC*, Section 501-5(f)].

14-33 What must be provided for switches, motor controllers, relays, fuses, or circuit breakers in Class I locations?
They must be provided with enclosures and must be approved for the location in which they are located [see *NEC*, Section 501-6(b)].

14-34 What is the required thickness of the sealing compound in Class I locations?
Not less than the trade size of the conduit, and in no case less than ⅝-inch thickness [see *NEC*, Section 501-5(c)(3)].

14-35 Must lighting fixtures for Class I locations be approved fixtures?
They definitely must be approved for the classification in which they are used, and must be so marked. Portable lamps must also be approved for these areas (see *NEC*, Section 501-9).

14-36 Is grounding necessary in Class I locations?

Yes, it is highly important. All exposed noncurrent-carrying parts of equipment are required to be grounded. Locknuts and bushings are not adequate grounding; they must have bonding jumpers around them. Where flexible conduit is used as permitted, bonding jumpers must be provided around such conduit (see *NEC*, Section 501-16).

14-37 Is surge protection required in Class I locations?

No, but if installed, lightning-protection devices are required on all ungrounded conductors. They must be connected ahead of the service-disconnecting means and must be bonded to the raceway at the service entrance (see *NEC*, Section 501-17).

14-38 What are some of the precautions that must be taken when transformers are used in Class II, Division 1 locations?

Transformers containing a flammable liquid must be installed only in approved vaults which are so constructed that a fire cannot be communicated to the hazardous area. Transformers that do not contain a flammable liquid must also be installed in vaults or must be enclosed in tight metal housings without ventilation [see *NEC*, Section 502-2(a)].

14-39 What type of wiring must be installed in Class II, Division 1 locations?

Rigid metal conduit (threaded), threaded steel intermediate metal conduit, or Type MI cable must be used. Boxes and fittings must have threaded bosses and tight-fitting covers and must be approved for the location [see *NEC*, Section 502-4(a)].

14-40 What type of wiring must be installed in Class II, Division 2 locations?

Rigid metal conduit, intermediate metal conduit, dust-tight wireways, electrical metallic tubing, or Type MI, MC, or SNM cable. Boxes, fittings, and joints must be made to minimize the entrance of dust [see *NEC*, Section 502-4(b)].

14-41 What is needed for sealing in Class II locations?

Where a raceway provides communication between an enclosure that is required to be dust-ignition-proof and one that is not, a suitable means must be provided to prevent the entrance of dust into the dust-ignition-proof enclosure through the raceway. This means may be a permanent and effective seal, a horizontal section of raceway not less than 10 feet long, or a vertical section of raceway not less than 5 feet long and extending downward from the dust-ignition-proof enclosure (see *NEC*, Section 502-5).

14-42 What type of fixtures must be provided for Class II locations?

The fixtures must be approved for these locations [see *NEC*, Section 502-11(a)].

14-43 Is there a difference between Class I fixtures and Class II fixtures?

Class I fixtures must be vaporproof and capable of withstanding and containing an explosion from within. Class I vapors usually have a higher flash point than dusts in a Class II location; therefore, the glass enclosure on Class I fixtures, while it must be heavier to withstand an explosion from within, may be smaller because of the higher flash temperatures. Class II fixtures are faced with a heat-dissipation problem because grain dusts and other types of dust have a low flashpoint temperature. Therefore, the glass enclosure on Class II fixtures must not be allowed to reach a high temperature.

14-44 Is grounding and bonding necessary in Class II locations?
Yes. All exposed noncurrent-carrying parts of equipment are required to be grounded. Locknuts and bushings are not adequate grounding; they must use bonding jumpers (see *NEC*, Section 502-16).

14-45 Is surge protection required in Class II locations?
If installed, lightning-protection devices of the proper type are required on all ungrounded conductors; they must be connected ahead of the service-disconnecting means and must be bonded to the raceway at the service entrance (see *NEC*, Section 502-17).

14-46 What type of wiring is required in Class III, Division 1 locations?
Threaded rigid metal conduit, threaded steel intermediate metal conduit, or Types MI, MC, SNM; boxes and fittings must have tight-fitting covers. There must not be any screw-mounting holes within the box through which sparks might escape [see *NEC*, Section 503-3(a)].

14-47 What vehicles are included under the commercial-garages classification?
These locations include places of storage, repairing, or servicing of self-propelled vehicles, including passenger automobiles, buses, trucks, tractors, etc., in which flammable liquids or flammable gases are used for fuel or power (see *NEC*, Section 511-1).

14-48 What are the various area classifications in commercial garages?
For each floor at or above grade, the entire area is considered as a Class I, Division 2 location to a level of 18 inches above the floor. When below grade, the entire area to 18 inches above grade is considered as a Class I, Division 2 location unless the area has positive ventilation, in which case it will be 18 inches above each such floor. Any pit or depression in the floor may be considered a Class I, Division 1 location by the enforcing authority. Adjacent stockrooms, etc., must be considered as Class I, Division 2 locations unless there is a tight curb or elevation of 18 inches above the hazardous area (see *NEC*, Section 511-3).

14-49 In commercial garages, how should areas adjacent to the hazardous area be classified?
Adjacent areas which by reason of ventilation, air pressure differentials, or physical spacing are such that in the opinion of the authority enforcing this Code no hazard exists shall be classified as nonhazardous [see *NEC*, Section 511-3(d)].

14-50 What material may be used in wiring spaces above Class I locations in commercial garages?
Rigid metal conduit, threaded steel intermediate metal conduit, rigid

nonmetallic conduit, electrical nonmetallic tubing, or Type MI, MC, or SNM cable [see *NEC*, Section 511-6(a)].

14-51 What is required for the 125-volt, single-phase, 15- and 20-ampere receptacles and protection of personnel in a commercial garage?

The receptacles are required to be protected by ground fault circuit interrupters, when they are used for portable lighting devices, electrical hand tools, and electrical automotive diagnostic equipment (see *NEC*, Section 511-10).

14-52 Where are seals required in commercial garages?

Wherever conduit passes from Class I, Division 2 areas to nonhazardous areas; this applies to horizontal as well as vertical boundaries of these areas (see *NEC*, Section 511-5).

14-53 What is required on equipment above the hazardous area in commercial garages?

Equipment that is less than 12 feet above the floor and that might produce sparks or particles of hot metal must be totally enclosed to prevent the escape of sparks or hot metal particles or be provided with guards or screens for the same purpose. Lighting fixtures must be located not less than 12 feet above the floor level unless protected by guards, screens, or covers (see *NEC*, Section 511-7).

14-54 What classification do pits or depressions of aircraft hangars require?

These are considered as Class I, Division 1 locations (see *NEC*, Section 513-2).

14-55 What is the normal classification of an aircraft hangar?

The entire area of a hangar, including adjacent areas not suitably cut off from the hangar, is considered as a Class I, Division 2 location to a height of 18 inches above the floor. Areas within 5 feet, horizontally, from aircraft power plants, aircraft fuel tanks, or aircraft structures containing fuel are considered as Class I, Division 2 locations to a point extending upward from the floor to a level 5 feet above the upper surfaces of wings and engine enclosures (see *NEC*, Section 513-2).

14-56 What type of wiring is required in all aircraft-hangar locations?

The wiring must be in accordance with Class I, Division 1 locations (see *NEC*, Section 513-3).

14-57 What type of wiring may be used in areas that are not considered as hazardous in aircraft hangars?

Metallic raceway, Type MI cable, TC, SNM, MC (see *NEC*, Section 513-4).

14-58 How must equipment, including lighting, above the aircraft be treated?

Any equipment or lighting less than 10 feet above wings and engine enclosures must be enclosed or suitably guarded to prevent the escape of arcs, sparks, or hot metal particles (see *NEC*, Section 513-5).

14-59 Is sealing required in aircraft hangars?

Yes, where wiring extends into or from hazardous areas. This regulation includes horizontal as well as vertical boundaries. Any raceways in the floor or below the floor are considered as being in the hazardous areas (see *NEC*, Section 513-7).

14-60 Is grounding required in aircraft hangars?

Yes, all raceways and all noncurrent-carrying metallic portions of fixed or portable equipment, regardless of voltage, must be grounded (see *NEC*, Section 513-12).

14-61 What is a gasoline dispensing and service station?

This includes locations where gasoline or other volatile flammable gases are transferred to the fuel tanks (including auxiliary fuel tanks) of self-propelled vehicles. Lubritoriums, service rooms, repair rooms, offices, sales-rooms, compressor rooms, and similar locations must conform to Article 511 of the *NEC* (see *NEC*, Section 514-1).

14-62 What are the hazardous areas in and around gasoline dispensing islands?

The area within the dispenser and extending for a distance of 18 inches in all directions from the enclosure and extending upward for a height of 4 feet from the driveway; any wiring within or below this area will be considered as a Class I, Division 1 location and must be approved for this location. Any area in an outside location within 20 feet, horizontally, from the exterior enclosure of any dispensing pump is considered as a Class I, Division 2 location. Any areas within buildings not suitably cut off from this 20-foot area must also be considered as a Class I, Division 2 location. The Class I, Division 2 location within the 20-foot area around dispensing pumps extends to a depth of 18 inches below the driveway or ground level. Any area in an outside location within 10 feet, horizontally, from any tank fill-pipe must be considered as a Class I, Division 2 location, and any area within a building in the 10-foot radius must be considered the same. Electrical wiring and equipment emerging from dispensing pumps must be considered as a Class I, Division 1 location, at least to its point of emergence from this area. Any area within a 3-foot radius of the point of discharge of any tank vent pipe must be considered as a Class I, Division 1 location; and below this point, the area must be considered as a Class I, Division 2 location. (See Table 514-2 of the *NEC*, which is very complete as to the classifications of all areas in and around service stations.)

14-63 What are the restrictions on switching circuits leading to or going through a gasoline dispensing pump?

The switches or circuit breakers in these circuits must be able to simultaneously disconnect all conductors of the circuit, including the grounded neutral, if any (see *NEC*, Section 514-5).

14-64 Are seals required in service station locations?

Yes, approved seals must be provided in each conduit run entering or leaving a gasoline dispensing pump or other enclosure located in a Class I, Division 1 or Division 2 location when connecting conduit originates in a nonhazardous location. The first fitting after the conduit emerges from the slab or from the concrete must be a sealing fitting (see *NEC*, Section 514-6).

14-65 Is grounding required in a service station wiring system?

Yes, all metal portions of dispensing islands, all metallic raceways, and all noncurrent-carrying parts of electrical equipment must be grounded, regardless of voltage (see *NEC*, Section 514-7).

14-66 What are bulk-storage plants?

Locations where gasoline or other volatile flammable liquids are stored in tanks having an aggregate capacity of one carload or more, and from which such products are distributed, usually by tank truck (see *NEC*, Section 515-1).

14-67 What are the classifications of bulk-storage plants?

Adequately ventilated indoor areas containing pumps, bleeders, withdrawal fittings, meters, and similar devices are considered as Class I, Division 2 locations in an area extending 5 feet in all directions from the exterior surfaces of such equipment; this location also extends 25 feet horizontally from any surface of the equipment and 3 feet above the floor or grade level. Areas that are not properly ventilated are required to have the same distances as above, but are considered as Class I, Division 1 locations (see *NEC*, Section 515-2).

14-68 Is underground wiring permitted in bulk-storage plants?

Yes, underground wiring must be installed in threaded rigid metal conduit. When buried in 2 feet or more of earth, it may be installed in rigid nonmetallic conduit or duct, or in the form of approved cable. Where cable is used, it must be enclosed in rigid metal conduit from the point of lowest buried cable level to the point of connection to the above-ground raceway (see *NEC*, Section 515-5).

14-69 What are spray application, dipping, and coating processes locations?

Locations where paints, lacquers, or other flammable finishes are used regularly or are frequently applied by spraying, dipping, brushing, or other means, and where readily ignitible deposits or residues from such paints, lacquers, or finishes may occur (see *NEC*, Section 516-1).

14-70 What is the classification of spray application, dipping, and coating processes locations?

They are considered as Class I, Division 1 locations (see *NEC*, Section 516-2).

14-71 Can a direct-drive fan be used for ventilation of spray application, dipping, and coating processes areas?

No, even though an explosionproof motor is used because the air passing over the motor contains particles of lacquers and paints. These particles will settle out on the motor, and the exterior temperature of the motor may reach a point at which these residues would become easily ignited (see *NEC*, Section 516-3).

14-72 Is there a lighting fixture approved for direct hanging in a spray-paint booth?

None have been approved because of the low flash point of residues that collect on the glass. Illumination through panels of glass or other translucent or transparent materials is permissible only where fixed lighting units are used. The panel effectively isolates the hazardous area from the area in which the lighting units are located, the lighting units are approved for their specific locations, and the panel is of material or is so protected that breakage will be unlikely. The arrangement is such that normal accumulations of hazardous residues on the surface of the panel will not be raised to a dangerous temperature by radiation or conduction from the illumination source (see *NEC*, Section 516-3).

14-73 What is a critical branch?

A subsystem of the emergency system consisting of feeders and branch circuits supplying energy to task illumination and selected receptacles serving areas and functions related to patient care, and which can be connected to alternate power sources by one or more transfer switches. (Definition in Section 517-2 of the *NEC*.)

14-74 What is a reference grounding point?

The terminal grounding bus which serves as the single focus for grounding the electrical equipment connected to an individual patient, or for grounding the metal or conductive furniture or other equipment within reach of the patient or a person who may be touching him. (Definition in Section 517-2 of the *NEC*.)

14-75 What is a room bonding point?

The terminal grounding bus which serves as a single focus for grounding the patient reference grounding buses and all other metal or conductive furniture, equipment, or structural surfaces in the room.

The bus may be located in or outside the room. The room reference grounding bus and the patient reference grounding bus may be a common bus if there is only one patient grounding bus in the room. (Definition in Section 517-2 of the *NEC*.)

14-76 What is the classification of a flammable anesthetics storage room?

It is considered as a Class I, Division 1 location throughout the entire area [see *NEC*, Section 517-60(a)(2)].

14-77 What classification is given to an anesthetizing location?

It is considered as a Class I, Division 1 location and this area extends upward to a height of 5 feet [see *NEC*, Section 517-60(a)(1)]. (Also see NFPA 99-1984.)

14-78 What type of wiring is required in a flammable-anesthetics location?

Any wiring operating at more than 8 volts between conductors must conform to Class I, Division 1 location wiring specifications and must be approved for the hazardous atmospheres involved (see *NEC*, Section 517-61)(a).

14-79 What special precautions pertain to circuits in anesthetizing areas?

All circuits must be supplied by approved isolating transformers, and proper ground-detector systems approved for the location must be used (see *NEC*, Section 517-61).

CHAPTER 15

Grounding
and
Ground Testing

15-1   What is considered the best grounding electrode?

Ten feet or more of buried metallic piping [see *NEC*, Section 250-81(a)]. At least one additional made electrode must be added for grounding to varied metallic water piping.

15-2   What is the maximum allowable resistance of made electrodes (grounding)?

The resistance to ground cannot exceed 25 ohms (see *NEC*, Section 250-84).

15-3   If the resistance exceeds 25 ohms, how may this condition be corrected?

By connecting two or more electrodes in parallel (see *NEC*, Section 250-84).

15-4   When driving a ground rod, or made electrode, is the resistance near the foundation more or less than the resistance of a rod driven a few feet away from the foundation?

The resistance will usually be more when a rod is driven near the foundation. This is due to the fact that when the rod is driven several feet away, the earth is in a circle around the ground rod instead of in a semicircle, as it would be when driven close to the foundation.

237

15-5 Is the ground resistance more or less when a rod is driven into undisturbed soil?

When a rod is driven into undisturbed soil, the pressure is greater against the rod. This pressure lowers the ground resistance.

15-6 Must all driven grounds be tested?

Since made electrodes must, where practical, have a resistance to ground not in excess of 25 ohms, the testing of ground resistance can be required by the inspection authority.

15-7 Is the common ohmmeter (DC-type) a good instrument to use when testing ground resistance?

No.

15-8 When using a common ohmmeter for ground testing, why can the results not be relied on?

Stray AC ground currents will probably be encountered. Some DC currents may also be found in the ground; these are due to electrolysis—the battery action between the moist earth and metals in contact with it, including the grounding electrode.

15-9 What instruments should be used for testing ground resistance?

A standard ground-testing megger, a battery-operated, vibrator-type ground tester, or a transistor-oscillator-type ground tester. The resistance may also be tested with an isolating transformer in conjunction with a voltmeter and an ammeter.

15-10 When testing the ground resistance on a ground rod, should the connection to the service be disconnected from the rod?

Yes.

15-11 Why should the connection to the ground rod be removed while testing?

Because of the danger of feedback into any other ground rods, equipment, etc., which would produce an inaccurate reading.

15-12 Sketch the procedure for ground-rod testing.

See Fig. 15-1.

15-13 Why is the middle electrode (B) in Fig. 15-1 set at 62% of the distance from A to C?

Referring to Fig. 15-1A, you will see that the knee of the resistance curve is at approximately 62%, and this point gives us the most accurate reading.

15-14 Sketch the procedure for testing ground resistance by the use of an isolating transformer, ammeter, and voltmeter.

See Fig. 15-2.

15-15 What other method may be used to lower the resistance of a ground rod besides paralleling rods?

The use of chemicals, such as magnesium sulfate, copper sulfate, or rock salt. While the use of various chemicals and salts will lower ground resistance, it will also corrode the grounding electrode. Because of this, the use of chemicals to lower grounding resistance is often frowned upon.

Fig. 15-1.   Ground-rod testing.

Fig. 15-2.   Distance of B from C measured electrode.

Fig. 15-3. Testing the ground resistance with an isolation transformer, ammeter, and voltmeter.

15-16 Approximately what quantity of chemicals is necessary for the treatment of soil to lower its ground resistance?
The first treatment should use from 40 to 90 pounds.

15-17 How is moisture added to the chemical treatment?
The normal amount of rainfall, in most places, will provide sufficient moisture. In extremely arid areas, the problem will be different.

15-18 Sketch two common methods of adding chemicals to the ground.
See Fig. 15-3.

Fig. 15-4. Adding chemicals to the soil to lower its ground resistance.

15-19 When paralleling ground rods, will the resistance be cut in proportion to the number of rods that are paralleled?

The resistance will not be cut proportionally. For instance, two rods paralleled and spaced 5 feet apart will cut the resistance to approximately 65% of the original resistance. Three rods in parallel and spaced 5 feet apart will cut the original resistance to approximately 42%. Four rods in parallel and spaced 5 feet apart will cut the original resistance to approximately 30%.

15-20 When testing the insulation resistance, which instrument will give the better results, an ohmmeter or a megger?

The megger will give the better results because it produces a higher output voltage, which often shows up defects that the low-voltage ohmmeter cannot indicate.

15-21 What type of equipment may be used in and around swimming pools?

All electrical equipment must be approved for this type of use (see *NEC*, Section 680-2).

15-22 What is the maximum voltage allowed for underwater lighting in a swimming pool?

No lighting can be operated at more than 150 volts [see *NEC*, Section 680-20(a)(2)].

15-23 May ground-fault circuit interrupters be used on underwater lighting?

Yes (see *NEC*, Sections 680-5 and 680-6).

15-24 May load-size conductors from ground-fault interrupters or transformers that supply underwater lighting be run with other electrical wiring or equipment?

No, they shall be kept independent [see *NEC*, Section 680-5(c)].

15-25 Must the noncurrent-carrying parts (metal) of lighting fixtures be grounded for the underwater lighting of swimming pools?

They must be grounded whether exposed or enclosed in nonconducting materials (see *NEC*, Sections 680-24 and 680-25).

15-26 How must underwater fixtures be installed in swimming pools?

Only approved fixtures may be installed, and they must be installed in the outside walls of the pool in closed recesses that are adequately drained and accessible for maintenance (see *NEC*, Section 680-20).

15-27 Do fixtures and fixture housings for underwater lighting have to be approved?

Yes, check in Underwriters Laboratories green book (see *NEC*, Section 680-20).

15-28 May galvanized rigid conduit be used to supply wet-niche fixtures?

No; conduit of brass or other corrosion-resistance metal must be used [see *NEC*, Section 680-20(b)].

15-29 May ordinary isolation transformers be used for supplying underwater lighting in swimming pools?

No; the transformer and enclosure must be identified for this purpose [see *NEC*, Section 680-5(a)].

15-30 How can approved equipment for swimming-pool usage be properly identified?

Look for the UL label; then look up the number and manufacturer's name in the UL listing books to ascertain whether it is listed for this type of use [see *NEC*, Section 110-3(b)].

15-31 How close to a swimming pool may an attachment-plug receptacle be installed?

Not closer than 10 feet from the inside walls of the swimming pool [see *NEC*, Section 680-6(a)].

15-32 Of what material must junction boxes that supply underwater pool lights be made?

Brass or other suitable copper alloy, suitable plastic, or other approved corrosion-resistance material, if less than 4 feet from the pool perimeter and less than 8 inches above the ground or concrete [see *NEC*, Section 680-21(a)].

15-33 How high or far from the pool shall approved transformers be mounted?

They are to be located not less than 4 inches, measured from the inside bottom of the enclosure to the ground level or pool deck, or not less than 8 inches to the maximum pool level, whichever provides the greatest elevation; also, not less than 4 feet from the inside wall of the pool unless separated from the pool by a solid fence, wall, or other permanent barrier (see *NEC*, Section 680-21).

15-34 What type of metal equipment must be grounded in pool areas?

All metallic conduit, piping systems, pool reinforcing steel, lighting fixtures, etc., must be bonded together and grounded to a common ground. This includes all metal parts of ladders, diving boards, and their supports (see *NEC*, Section 680-24).

15-35 What is considered adequate in the bonding of the reinforcing bar in concrete?

Tying the reinforcing bar together with wire, as is customarily done, is considered adequate, provided the job is well done. Welding, of course, would also be acceptable [see *NEC*, Section 680-22(a), Exception No. 1].

15-36 May the electrical equipment be grounded to a separate grounding electrode on a swimming-pool insulation?

No, the grounding must be common to the deck box or transformer ground (see *NEC*, Section 680-25).

15-37 What is the minimum size of grounding conductor permitted for deck boxes on swimming pools?

The equipment grounding conductor must be No. 12 AWG or larger [see *NEC*, Section 680-25(b)(1)].

15-38 May metal raceways be relied on for grounding in swimming-pool areas?

No, an insulated equipment grounding conductor is required [see *NEC*, Section 680-25(b)(1)].

15-39 How must pumps, water-treating equipment, etc., in swimming-pool areas be bonded?

They shall be bonded by a solid copper conductor not smaller than No. 8 AWG [see *NEC*, Section 680-22(b)].

15-40 May structural reinforcing steel be used for bonding nonelectrical parts in swimming-pool areas?

Yes, where the requirements of Section 250-113 are met and reliable and approved connections can be made [see *NEC*, Section 680-22(a), Exception No. 2].

15-41 What are the clearances for service-drop conductors in swimming-pool locations?

They must be installed not less than 10 feet horizontally from the pool edge, diving structures, observation stands, towers, or platforms, and must not be installed above the swimming pool or surrounding area within the 10-foot area (see *NEC*, Section 680-8).

15-42 What type of wiring is required in theaters and assembly halls?

The wiring must be metal raceways or Type MC or MI cable, with some exceptions (see *NEC*, Section 518-4 and 520-4).

15-43 Where the assembly area is less than a 100-person capacity, may the type of wiring be altered?

Yes, Type AC metal-clad, knob-and-tube, or Type NM cable may be used. However, it is not considered good practice to use anything other than metal raceways. Also, most local ordinances or laws prohibit the use of anything other than metal raceways in public places of assembly (see *NEC*, Section 518-1).

15-44 How may the population capacity be determined in places of assembly?

Refer to the *NFPA Life Safety Code (No. 101)*.

15-45 What is fire-rated construction in a place of assembly?

The fire resistive classification used in building codes (see *NEC*, Section 518-4, FPN).

15-46 What are the fixed wiring methods recognized for use in a theater or similar location?

Metal raceways, nonmetallic raceways encased in at least 2 inches of concrete, Type MI cable, or Type MC cable (see *NEC*, Section 520-4).

The answers to questions 15-47 to 15-77 may be found on pages 247–248.

15-47 In the diagram (Fig. 15-5) of a 120/240-volt, three-wire circuit, load A is 10 amperes and load B is 5 amperes. Will the neutral carry any current, and, if so, how much will it carry?

(a) No current
(b) 10 amperes
(c) 15 amperes
(d) 5 amperes

Fig. 15-5. The neutral lead carries 5 amperes of current.

15-48 In the 120/240-volt, three-wire circuit (Fig. 15-5), the neutral is open at point X. The resistance of load A is 10 ohms; load B, 12 ohms; load C, 24 ohms; and load D, 20 ohms. What is the voltage drop of loads B and C, respectively?
(a) B = 160 volts; C = 80 volts
(b) B = 80 volts; C = 160 volts
(c) B = 120 volts; C = 120 volts
(d) B = 240 volts; C = 240 volts

Fig. 15-6. With the neutral lead open at point X, the voltage drop of load B is 80 volts, and the voltage drop of load C is 160 volts.

15-49 If the neutral is No. 10 wire or larger, what appliance, if any, may use the neutral as an equipment ground?
(a) None
(b) Grounding-type receptacle
(c) Electric motor
(d) Electric range (except when the range is fed from a feeder panel)

15-50 When pulling current-carrying conductors into conduits, at what number of such conductors does derating of conductor current-carrying capacity begin?

(a) Four
(b) Six
(c) No derating
(d) Three

15-51 Under what condition, if any, is it possible or permissible to put overload protection in a neutral conductor?

(a) Always
(b) Never permissible
(c) Where the overload device simultaneously opens all conductors

15-52 Under what conditions may a yellow neutral be used?
(a) When extra color coding is advantageous
(b) Never
(c) On rewire work

15-53 The total load on a circuit for air conditioning only cannot exceed what percentage of the branch-circuit load?
(a) 115%
(b) 75%
(c) 80%

15-54 The total load of an air conditioner cannot exceed what percentage of the branch-circuit load on a circuit that also supplies lighting?
(a) 50%
(b) 75%
(c) 80%

15-55 Hazardous concentrations of gases or vapors are not normally present in Class I, Division 1 locations.
T F (True or False)

15-56 A location where cloth is woven should be designated as Class III, Division 1.
T F

15-57 Resistance- and reactive-type dimmers may be placed in the ungrounded conductor.
T F

15-58 With outline lighting, stranded conductors need not be soldered when pin-type sockets are used.
T F

15-59 A separate connection from the service drop is never acceptable as an emergency service installation.
T F

15-60 Portable, explosionproof lighting units may be employed inside a spray booth during operation.
T F

15-61 A plug receptacle exclusively for the janitor's use may not be tapped from the emergency circuit wiring.
T F

15-62 Type NM (nonmetallic) cable may be used for under-plaster extensions.
T F

15-63 Heating cable may not be used with metal lath.
T F

15-64 Time switches need not be of the externally operated type.
T F

15-65 Heating cables in a concrete floor must be placed on at least 2-inch centers.
T F

15-66 Double-throw knife switches may be mounted either vertically or horizontally.
T F

15-67 Outlets for heavy-duty lampholders must be rated at a minimum load of
(a) 600 volt-amperes
(b) 1320 volt-amperes
(c) 2 amperes
(d) 5 amperes

15-68 Where permissible, the demand factor applied to that portion of the unbalanced neutral load in excess of 200 amperes is
(a) 40%
(b) 80%
(c) 70%
(d) 60%

15-69 Fuses must never be connected in multiple.
T F

15-70 No. 18 copper wire may be employed for grounding a portable device used on a 20-ampere circuit.
T F

15-71 Secondary circuits of wound-rotor induction motors require separate overcurrent protection.
T F

15-72 A neutral conductor is sometimes smaller than the ungrounded conductors.
T F

15-73 The largest size conductor permitted in underfloor raceways is
(a) largest that raceway is designed for
(b) No. 0
(c) 250 MCM
(d) No. 000

15-74 The maximum rating of a portable appliance used on a 20-ampere branch circuit must be
(a) 16 amperes
(b) 10 amperes
(c) 20 amperes
(d) 12 amperes

15-75 The demand factor that may be applied to the neutral of an electric range is
(a) 40%
(b) 80%
(c) 70%
(d) 60%

15-76 The unprotected length of a tap conductor having a current-carrying capacity equal to one-third that of the main conductor must be no greater than
(a) 15 feet
(b) 25 feet
(c) 20 feet
(d) 10 feet

15-77 The minimum feeder allowance for show-window lighting expressed in volt-amperes per linear foot shall be
(a) 100 volt-amperes
(b) 200 volt-amperes
(c) 300 volt-amperes
(d) 500 volt-amperes

Answers to Questions 15-47 to 15-77

15-47 d
15-48 b
15-49 d
15-50 a
15-51 c
15-52 b
15-53 c [see *NEC*, Section 440-62(b)]
15-54 a [see *NEC*, Section 440-62(c)]
15-55 F [see *NEC*, Section 500-5(a)]
15-56 T [see *NEC*, Section 500-7(a)]

15-57 T [see *NEC*, Section 520-25(b)]
15-58 T [see *NEC*, Section 600-21(e)]
15-59 F [see *NEC*, Section 700-12(d)]
15-60 F [see *NEC*, Section 516-3]
15-61 T [see *NEC*, Section 700-15]
15-62 F [see *NEC*, Section 344-2]
15-63 F [see *NEC*, Section 424-41(c)]
15-64 T [see *NEC*, Section 380-5]
15-65 F [see *NEC*, Section 426-20(b)]
15-66 T [see *NEC*, Section 380-6(b)]
15-67 a [see *NEC*, Section 220-3(c)]
15-68 c [see *NEC*, Section 220-22]
15-69 T [see *NEC*, Section 240-8, for the one exception]
15-70 T [see *NEC*, Section 250-95, Exception No. 1]
15-71 F [see *NEC*, Section 430-32(d)]
15-72 T [see *NEC*, Section 220-22]
15-73 a [see *NEC*, Section 354-4]
15-74 a [see *NEC*, Section 210-22(c)]
15-75 c [see *NEC*, Section 220-22]
15-76 b [see *NEC*, Section 240-21, Exception No. 3]
15-77 b [see *NEC*, Section 220-12]

15-78 All 125-volt, single-phase, 15- and 20-ampere dwelling unit receptacle outlets, installed above countertops within 6 feet of a kitchen sink, require ground-fault, circuit interrupter protection.
True [see *NEC*, Section 210-8(a)].

15-79 Cable assemblies with a neutral conductor sized smaller than the ungrounded conductors require no marking.
False [see *NEC*, Section 210-19(a)].

15-80 Type AC cable (BX) installed in thermal insulation is required to be a cable assembly with 90° C-rated conductors.
True (see *NEC*, Section 333-5, Exception).

15-81 It is not necessary to bury lengths of rigid or intermediate metal conduit 24 inches below a driveway, street, alley, or road.
False [see *NEC*, Section 300-5(a), Exception No. 3].

15-82 Number 16 and 18 AWG fixture wires are required to be counted when sizing a junction box with a combination of other branch-circuit conductors.
True [see *NEC*, Section 370-6(a)].

15-83 A second set of insulated equipment grounding conductors, installed in a box, is required to be counted for box fill.
True [see *NEC*, Section 370-6(a)].

15-84 Galvanized steel wires are permitted for the securing of a race-way, cable assembly, or box above a T-Bar ceiling.

True, when they provide a rigid support [see *NEC*, Sections 300-11(a) and 370-13(b)].

15-85 A disconnecting means for an HVAC system can be installed within an air-conditioning or refrigeration equipment enclosure, when a tool is required to remove the cover to gain access.
False (see *NEC* Definition of Accessible [equipment] and Section 440-14).

15-86 Identification is not required for emergency circuit wiring boxes or enclosures.
False (see *NEC*, Section 700-9).

15-87 Cable trays are permitted to be used for Type SE cable feeders in an apartment building or commercial location.
True [see *NEC*, Section 318-3(b), FPN].

15-88 An electric baseboard heater can be installed beneath a dwelling unit receptacle outlet.
False [see *NEC*, Section 110-3(b) and 210-52(a), FPN].

15-89 All dwelling unit basement receptacles are required to be protected by a GFCI.
True [see *NEC*, Section 210-8(a)(4)].

15-90 Floating buildings are required to conform to the *National Electrical Code*.
True (see *NEC*, Article 553).

15-91 A grounded service-entrance conductor used in a three-wire, single-phase dwelling service or feeder can be reduced in size.
True (see *NEC*, Article 310, Note 3 to Tables 310-16 through 310-19).

15-92 Welding cables are not permitted to be installed in a cable tray.
False [see *NEC*, Section 318-3(b)(1), Exception].

15-93 GFCI protection is not required for the receptacles in a repair garage.
False (see *NEC*, Section 511-10).

15-94 GFCI protection is required for a portable high-pressure sprayer.
True [see *NEC*, Section 422-8(d)(3)].

15-95 Electric meter socket enclosures are covered by the *National Electrical Code*.
True (see *NEC*, Section 373-1).

15-96 Type SE cable assemblies, used for a service, are permitted to be located next to a window, when the drip loop cannot be contacted from the sides, bottom, or top of a window.
True (see *NEC*, Sections 230-9 and 338-1).

15-97 The *Code* recognizes the use of a raceway for enclosing a combination of systems, such as water, gas, or air lines.
False [see *NEC*, Section 300-8 and the FPNs to Sections 356-2, 511-6(a), and 516-7(a)].

15-98 Flexible metal conduit is not permitted for use as a service-entrance raceway.

False [see *NEC*, Section 230-43(1)].

15-99 Hospital-grade receptacles are not required within any other area of a hospital, other than within an inhalation anesthetizing location.

False [see *NEC*, Section 517-18(b)].

15-100 Nonmetallic sheathed cables (romex) require protection in the form of sleeves or grommets, prior to being installed through the sharp edges of a metal stud partition.

True [see *NEC*, Section 300-4(b)(1)].

15-101 Ceiling paddle fans are required to be supported by a structural member, instead of the box used at the fan outlet.

True [see *NEC*, Section 370-17(c)].

15-102 Where electronic computer/data processing equipment is installed in a room, one disconnecting means can serve to disconnect the power and air-conditioning systems.

True (see *NEC*, Section 645-10).

CHAPTER 16

High-tech Wiring

While virtually all of the work that electrical contractors perform has been 60 Hz wiring for light and power, this is no longer so, and is destined to be even less so in the future. Certainly, wiring for light and power will not go away, but it will comprise a far lower percentage of the work electrical installers perform.

New types of technologies are coming into almost every business and industry. Computer technology, along with stunning advances in genetic engineering and biotechnology, is drastically changing our world; another set of technologies is required to support these advances. These supporting technologies are going to be in high demand in the next several decades. Both individuals and businesses will need them to use the newfound advantages of the technological age.

As we mentioned in the preface, you should begin developing your high-tech skills immediately. The electronics-based technologies have grown from a small percentage of the total volume of electrical contracting in 1980 to roughly one third today. This segment of our trade is growing faster than we thought it would, and is becoming very important. You should put more of your money and effort into learning about these technologies (how to install them, how to estimate them, how to sell them) than on anything else. They are the future.

16-1 Why are optical fiber cables preferable to copper conductors for communications?

Because they can carry many times as many signals, with far less degradation of the signals. One pair of optical fiber cables can carry 1,344 telephone conversations.

16-2 What are the three basic parts of optical fiber cables?

The core, the cladding, and the outer covering.

16-3 What are the two most important things to remember when installing optical fiber cables?

First, that the pulling force should not be applied to the optical fiber, but to a "strength member." Second, that optical fiber cables cannot be sharply bent.

16-4 What is the most difficult operation of connecting optical fiber cables?

Terminating or splicing them. This can be a time-consuming and difficult process.

16-5 Optical fiber cables do not carry electricity. Why are they covered within the *NEC*?

For two reasons: First, because they are often installed with electrical cables, and by the same installers. Second, because they are dependent upon electrical and electronic systems, and are connected to them.

16-6 Does the *NEC* cover all optical fiber installations?

No, it only applies to the types of installations that are covered in Article 770 of the *NEC* (see *NEC*, Section 770-1).

16-7 Can optical fiber cables be installed in cable trays?

Yes (see *NEC*, Section 770-5).

16-8 What types of optical fiber cables are allowed in plenums?

Types OFNP and OFPC, OFNR, OFCR, OFN, and OFC may also be installed in plenums under certain circumstances [see *NEC*, Section 770-53(a)].

16-9 What does the term photovoltaic mean?

Generating electricity from light.

16-10 What is a photovoltaic cell?

A special type of semiconductor that generates a small voltage when exposed to light.

16-11 What type of electricity do photovoltaic cells generate?

Direct current only.

16-12 What is an array?

An assembly of photovoltaic panels (see *NEC*, Section 690-1).

16-13 Why are blocking diodes required for photovoltaic systems?

To prevent battery currents from flowing backward through the photovoltaic cells when no sunlight is present.

16-14 Do photovoltaic systems require special disconnects?
Yes. A disconnecting means must be installed that will disconnect photovoltaic conductors from all other system conductors (see *NEC*, Section 690-15).

16-15 Can any recognized type of wiring be used for photovoltaic systems?
Yes [see *NEC*, Section 690-31(a)].

16-16 Is it required that a 12-volt photovoltaic system be grounded?
No (see *NEC*, Section 690-41).

16-17 Name two types of communications signals.
Analog and digital.

16-18 In general, what type of signal is better?
Digital; it provides a much clearer and stronger signal.

16-19 What is a protector?
Protectors are surge-suppressor-type devices. They are required for many outdoor communication circuits (see *NEC*, Section 800-30).

16-20 What is the smallest grounding conductor that can be used for communication circuits?
Number 14 copper [see *NEC*, Section 800-40(a)(3)].

16-21 What steps must be taken when communication grounding conductors are installed in metal conduit?
The metal conduit must be bonded on both ends [see *NEC*, Section 800-40(a)(5)].

16-22 Should communication grounds be bonded to power system grounds?
Yes [see *NEC*, Section 800-40(d)].

16-23 How far must communication circuits be separated from power circuits?
At least 2 inches [see *NEC*, Section 800-52(a)(1)].

16-24 In what part of the *NEC* would you find requirements for fire alarm systems and circuits?
In Article 760.

16-25 What is a closed-loop system?
A special type of power distribution that is controlled jointly by signals from the controlling equipment and the utilization equipment.

16-26 Where are closed-loop power systems intended to be used?
In the Smart House™ being developed by the National Association of Home Builders.

Index

AUDEL®

Over a Century of Excellence
for the Professional
and
Vocational Trades and the Crafts

Order now from your local bookstore
or use the convenient order form
at the back of this book.

AUDEL

These fully illustrated, up-to-date guides and manuals mean a better job done for mechanics, engineers, electricians, plumbers, carpenters, and all skilled workers.

ELECTRICAL

House Wiring (Seventh Edition)
ROLAND E. PALMQUIST;
revised by PAUL ROSENBERG

*5 1/2 × 8 1/4 Hardcover 248 pp. 150 Illus.
ISBN: 0-02-594692-7 $22.95*

Rules and regulations of the current 1990 National Electrical Code for residential wiring fully explained and illustrated.

Practical Electricity
(Fifth Edition)
ROBERT G. MIDDLETON;
revised by L. DONALD MEYERS

*5 1/2 × 8 1/4 Hardcover 512 pp. 335 Illus.
ISBN: 0-02-584561-6 $19.95*

The fundamentals of electricity for electrical workers, apprentices, and others requiring concise information about electric principles and their practical applications.

Guide to the 1993 National Electrical Code
ROLAND E. PALMQUIST;
revised by PAUL ROSENBERG

*5 1/2 × 8 1/4 Paperback 608 pp.
100 line drawings
ISBN: 0-02-077761-2 $25.00*

The guide to the most recent revision of the electrical codes—how to read them, under-

stand them, and use them. Here is the most authoritative reference available, making clear the changes in the code and explaining these changes in a way that is easy to understand.

Installation Requirements of the 1993 National Electrical Code
PAUL ROSENBERG

*5 1/2 × 8 1/4 Paperback 261 pp.
100 line drawings
ISBN: 0-02-077760-4 $22.00*

A handy guide for electricians, contractors, and architects who need a reference on location. Arranging all the pertinent requirements (and only the pertinent requirements) of the 1993 NEC, it has an easy-to-follow format. Concise and updated, it's a perfect working companion for Apprentices, Journeymen, or for Master electricians.

Mathematics for Electricians and Electronics Technicians
REX MILLER

*5 1/2 × 8 1/4 Hardcover 312 pp. 115 Illus.
ISBN: 0-8161-1700-4 $14.95*

Mathematical concepts, formulas, and problem-solving techniques utilized on-the-job by electricians and those in electronics and related fields.

Fractional-Horsepower Electric Motors

REX MILLER and
MARK RICHARD MILLER

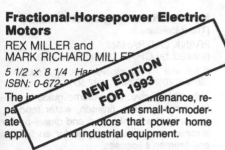

5 1/2 × 8 1/4 Ha~
ISBN: 0-672-~
The i~ ~tenance, re-
pa~ ~e small-to-moder-
ate~ ~otors that power home
appl~ ~d industrial equipment.

Electric Motors (Fifth Edition)

EDWIN P. ANDERSON
and REX MILLER

5 1/2 × 8 1/4 Hardcover 696 pp.
Photos/line art
ISBN: 0-02-501920-1 $35.00

Complete reference guide for electricians, industrial maintenance personnel, and installers. Contains both theoretical and practical descriptions.

Home Appliance Servicing
(Fourth Edition)

EDWIN P. ANDERSON;
revised by REX MILLER

5 1/2 × 8 1/4 Hardcover 640 pp. 345 Illus.
ISBN: 0-672-23379-7 $22.50

The essentials of testing, maintaining, and repairing all types of home appliances.

Television Service Manual
(Fifth Edition)

ROBERT G. MIDDLETON;
revised by JOSEPH G. BARRILE

5 1/2 × 8 1/4 Hardcover 512 pp. 395 Illus.
ISBN: 0-672-23395-9 $16.95

A guide to all aspects of television transmission and reception, including the operating principles of black and white and color receivers. Step-by-step maintenance and repair procedures.

Electrical Course for Apprentices and Journeymen
(Third Edition)

ROLAND E. PALMQUIST

5 1/2 × 8 1/4 Hardcover 478 pp. 290 Illus.
ISBN: 0-02-594550-5 $19.95

This practical course in electricity for those in formal training programs or learning on their own provides a thorough understanding of operational theory and its applications on the job.

Questions and Answers for Electricians Examinations
(1993 NEC Rulings Included)

revised by PAUL ROSENBERG

5 1/2 × 8 1/4 Paperback 270 pp.
100 line drawings
ISBN: 0-02-077762-0 $20.00

An Audel classic, considered the most thorough work on the subject in coverage and content. This fully revised edition is based on the 1993 National Electrical Code®, and is written for anyone preparing for the various electricians' examinations—Apprentice, Journeyman, or Master. It provides the license applicant with an understanding of theory as well as of all definitions, specifications, and regulations included in the new NEC.

MACHINE SHOP AND MECHANICAL TRADES

Machinists Library
(Fourth Edition, 3 Vols.)

REX MILLER

5 1/2 × 8 1/4 Hardcover 1,352 pp. 1120 Illus.
ISBN: 0-672-23380-0 $52.95

An indispensable three-volume reference set for machinists, tool and die makers, machine operators, metal workers, and those with home workshops. The principles and methods of the entire field are covered in an up-to-date text, photographs, diagrams, and tables.

Volume I: Basic Machine Shop
REX MILLER

5 1/2 × 8 1/4 Hardcover 392 pp. 375 Illus.
ISBN: 0-672-23381-9 $17.95

Volume II: Machine Shop
REX MILLER

5 1/2 × 8 1/4 Hardcover 528 pp. 445 Illus.
ISBN: 0-672-23382-7 $19.95

Volume III: Toolmakers Handy Book
REX MILLER

5 1/2 × 8 1/4 Hardcover 432 pp. 300 Illus.
ISBN: 0-672-23383-5 $14.95

Mathematics for Mechanical Technicians and Technologists
JOHN D. BIES

5 1/2 × 8 1/4 Hardcover 342 pp. 190 Illus.
ISBN: 0-02-510620-1 $17.95

The mathematical concepts, formulas, and problem-solving techniques utilized on the job by engineers, technicians, and other workers in industrial and mechanical technology and related fields.

Millwrights and Mechanics
Guide (Fourth Edition)
CARL A. NELSON

5 1/2 × 8 1/4 Hardcover 1,040 pp. 880 Illus.
ISBN: 0-02-588591-x $29.95

The most comprehensive and authoritative guide available for millwrights, mechanics, maintenance workers, riggers, shop workers, foremen, inspectors, and superintendents on plant installation, operation, and maintenance.

Welders Guide (Third Edition)
JAMES E. BRUMBAUGH

5 1/2 × 8 1/4 Hardcover 960 pp. 615 Illus.
ISBN: 0-672-23374-6 $23.95

The theory, operation, and maintenance of all welding machines. Covers gas welding equipment, supplies, and process; arc welding equipment, supplies, and process; TIG and MIG welding; and much more.

Welders/Fitters Guide
HARRY L. STEWART

8 1/2 × 11 Paperback 160 pp. 195 Illus.
ISBN: 0-672-23325-8 $7.95

Step-by-step instruction for those training to become welders/fitters who have some knowledge of welding and the ability to read blueprints.

Sheet Metal Work
JOHN D. BIES

5 1/2 × 8 1/4 Hardcover 456 pp. 215 Illus.
ISBN: 0-8161-1706-3 $19.95

An on-the-job guide for workers in the manufacturing and construction industries and for those with home workshops. All facets of sheet metal work detailed and illustrated by drawings, photographs, and tables.

Power Plant Engineers Guide
(Third Edition)
FRANK D. GRAHAM;
revised by CHARLIE BUFFINGTON

5 1/2 × 8 1/4 Hardcover 960 pp. 530 Illus.
ISBN: 0-672-23329-0 $27.50

This all-inclusive, one-volume guide is perfect for engineers, firemen, water tenders, oilers, operators of steam and diesel-power engines, and those applying for engineer's and firemen's licenses.

Mechanical Trades Pocket
Manual (Third Edition)
CARL A. NELSON

4 × 6 Paperback 364 pp. 255 Illus.
ISBN: 0-02-588665-7 $14.95

A handbook for workers in the industrial and mechanical trades on methods, tools, equipment, and procedures. Pocket-sized for easy reference and fully illustrated.

PLUMBING

Plumbers and Pipe Fitters
Library (Fourth Edition, 3 Vols.)
CHARLES N. McCONNELL

5 1/2 × 8 1/4 Hardcover 952 pp. 560 Illus.
ISBN: 0-02-582914-9 $68.45

This comprehensive three-volume set contains the most up-to-date information available for master plumbers, journeymen, apprentices, engineers, and those in the building trades. A detailed text and clear diagrams, photographs, and charts and tables treat all aspects of the plumbing, heating, and air conditioning trades.

Volume I: Materials, Tools, Roughing-In
CHARLES N. McCONNELL;
revised by TOM PHILBIN

5 1/2 × 8 1/4 Hardcover 304 pp. 240 Illus.
ISBN: 0-02-582911-4 $20.95

Volume II: Welding, Heating, Air Conditioning
CHARLES N. McCONNELL;
revised by TOM PHILBIN

5 1/2 × 8 1/4 Hardcover 384 pp. 220 Illus.
ISBN: 0-02-582912-2 $22.95

Volume III: Water Supply, Drainage, Calculations
CHARLES N. McCONNELL;
revised by TOM PHILBIN

5 1/2 × 8 1/4 Hardcover 264 pp. 100 Illus.
ISBN: 0-02-582913-0 $20.95

The Home Plumbing Handbook
(Fourth Edition)

CHARLES N. McCONNELL

8 1/2 × 11 Paperback 224 pp. 210 Illus.
ISBN: 0-02-079651-X $17.00

This handy, thorough volume, a longtime standard in the field with the professional, has been updated to appeal to the do-it-yourself plumber. Aided by the book's many illustrations and manufacturers' instructions, the home plumber is guided through most basic plumbing procedures. All techniques and products conform to the latest changes in codes and regulations.

The Plumbers Handbook
(Eighth Edition)

JOSEPH P. ALMOND, SR.;
revised by REX MILLER

4 × 6 Paperback 368 pp. 170 Illus.
ISBN: 0-02-501570-2 $19.95

Comprehensive and handy guide for plumbers and pipefitters—fits in the toolbox or pocket. For apprentices, journeymen, or experts.

Questions and Answers for Plumbers' Examinations
(Third Edition)

JULES ORAVETZ;
revised by REX MILLER

5 1/2 × 8 1/4 Paperback 288 pp. 145 Illus.
ISBN: 0-02-593510-0 $14.95

Complete guide to preparation for the plumbers' exams given by local licensing authorities. Includes requirements of the National Bureau of Standards.

HVAC

Air Conditioning: Home and Commercial (Fourth Edition)

EDWIN P. ANDERSON;
revised by REX MILLER

5 1/2 × 8 1/4 Hardcover 528 pp. 180 Illus.
ISBN: 0-02-584885-2 $29.95

A guide to the construction, installation, operation, maintenance, and repair of home, commercial, and industrial air conditioning systems.

Heating, Ventilating, and Air Conditioning Library
(Second Edition, 3 Vols.)

JAMES E. BRUMBAUGH

5 1/2 × 8 1/4 Hardcover 1,840 pp. 1,275 Illus.
ISBN: 0-672-23388-6 $53.85

An authoritative three-volume reference library for those who install, operate, maintain, and repair HVAC equipment commercially, industrially, or at home.

Volume I: Heating Fundamentals, Furnaces, Boilers, Boiler Conversions

JAMES E. BRUMBAUGH

5 1/2 × 8 1/4 Hardcover 656 pp. 405 Illus.
ISBN: 0-672-23389-4 $17.95

Volume II: Oil, Gas and Coal Burners, Controls, Ducts, Piping, Valves

JAMES E. BRUMBAUGH

5 1/2 × 8 1/4 Hardcover 592 pp. 455 Illus.
ISBN: 0-672-23390-8 $17.95

Volume III: Radiant Heating, Water Heaters, Ventilation, Air Conditioning, Heat Pumps, Air Cleaners

JAMES E. BRUMBAUGH

5 1/2 × 8 1/4 Hardcover 592 pp. 415 Illus.
ISBN: 0-672-23391-6 $17.95

Oil Burners (Fifth Edition)

EDWIN M. FIELD

5 1/2 × 8 1/4 Hardcover 360 pp. 170 Illus.
ISBN: 0-02-537745-0 $29.95

An up-to-date sourcebook on the construction, installation, operation, testing, servicing, and repair of all types of oil burners, both industrial and domestic.

Refrigeration: Home and Commercial (Fourth Edition)

EDWIN P. ANDERSON;
revised by REX MILLER

5 1/2 × 8 1/4 Hardcover 768 pp. 285 Illus.
ISBN: 0-02-584875-5 $34.95

A reference for technicians, plant engineers, and the homeowner on the installation, operation, servicing, and repair of everything from single refrigeration units to commercial and industrial systems.

PNEUMATICS AND HYDRAULICS

Hydraulics for Off-the-Road Equipment (Second Edition)

HARRY L. STEWART;
revised by TOM PHILBIN

5 1/2 × 8 1/4 Hardcover 256 pp. 175 Illus.
ISBN: 0-8161-1701-2 $13.95

This complete reference manual on heavy equipment covers hydraulic pumps, accumulators, and motors; force components; hydraulic control components; filters and filtration, lines and fittings, and fluids; hydrostatic transmissions; maintenance; and troubleshooting.

Pneumatics and Hydraulics (Fourth Edition)

HARRY L. STEWART;
revised by TOM STEWART

5 1/2 × 8 1/4 Hardcover 512 pp. 315 Illus.
ISBN: 0-672-23412-2 $19.95

The principles and applications of fluid power. Covers pressure, work, and power; general features of machines; hydraulic and pneumatic symbols; pressure boosters; air compressors and accessories; and much more.

Pumps (Fifth Edition)

HARRY L. STEWART;
revised by REX MILLER

5 1/2 × 8 1/4 Hardcover 552 pp. 360 Illus.
ISBN: 0-02-614725-4 $35.00

The practical guide to operating principles of pumps, controls, and hydraulics. Covers installation and day-to-day service.

CARPENTRY AND CONSTRUCTION

Carpenters and Builders Library (Sixth Edition, 4 Vols.)

JOHN E. BALL;
revised by JOHN LEEKE

5 1/2 × 8 1/4 Hardcover 1,300 pp. 988 Illus.
ISBN: 0-02-506455-4 $89.95

This comprehensive four-volume library has set the professional standard for decades for carpenters, joiners, and woodworkers.

Volume 1: Tools, Steel Square, Joinery
JOHN E. BALL;
revised by JOHN LEEKE

5 1/2 × 8 1/4 Hardcover 377 pp. 340 Illus.
ISBN: 0-02-506451-7 $21.95

Volume 2: Builders Math, Plans, Specifications
JOHN E. BALL;
revised by JOHN LEEKE

5 1/2 × 8 1/4 Hardcover 319 pp. 200 Illus.
ISBN: 0-02-506452-5 $21.95

Volume 3: Layouts, Foundation, Framing
JOHN E. BALL;
revised by JOHN LEEKE

5 1/2 × 8 1/4 Hardcover 269 pp. 204 Illus.
ISBN: 0-02-506453-3 $21.95

Volume 4: Millwork, Power Tools, Painting
JOHN E. BALL;
revised by JOHN LEEKE

5 1/2 × 8 1/4 Hardcover 335 pp. 244 Illus.
ISBN: 0-02-506454-1 $21.95

Complete Building Construction (Second Edition)

JOHN PHELPS;
revised by TOM PHILBIN

5 1/2 × 8 1/4 Hardcover 744 pp. 645 Illus.
ISBN: 0-672-23377-0 $22.50

Constructing a frame or brick building from the footings to the ridge. Whether the building project is a tool shed, garage, or a complete home, this single fully illustrated volume provides all the necessary information.

Complete Roofing Handbook (Second Edition)

JAMES E. BRUMBAUGH
revised by JOHN LEEKE

5 1/2 × 8 1/4 Hardcover 536 pp. 510 Illus.
ISBN: 0-02-517851-2 $30.00

Covers types of roofs; roofing and reroofing; roof and attic insulation and ventilation; skylights and roof openings; dormer construction; roof flashing details; and much more. Contains new information on code requirements, underlaying, and attic ventilation.

Complete Siding Handbook (Second Edition)

JAMES E. BRUMBAUGH
revised by JOHN LEEKE

5 1/2 × 8 1/4 Hardcover 440 pp. 320 Illus.
ISBN: 0-02-517881-4 $30.00

This companion volume to the *Complete Roofing Handbook* has been updated to reflect current emphasis on compliance with building codes. Contains new sections on spunbound olefin, building papers, and insulation materials other than fiberglass.

Masons and Builders Library
(Second Edition, 2 Vols.)
LOUIS M. DEZETTEL;
revised by TOM PHILBIN

5 1/2 × 8 1/4 Hardcover 688 pp. 500 Illus.
ISBN: 0-672-23401-7 $27.95

This two-volume set provides practical instruction in bricklaying and masonry. Covers brick; mortar; tools; bonding; corners, openings, and arches; chimneys and fireplaces; structural clay tile and glass block; brick walls; and much more.

Volume 1: Concrete, Block, Tile, Terrazzo
LOUIS M. DEZETTEL;
revised by TOM PHILBIN

5 1/2 × 8 1/4 Hardcover 304 pp. 190 Illus.
ISBN: 0-672-23402-5 $14.95

Volume 2: Bricklaying, Plastering, Rock Masonry, Clay Tile
LOUIS M. DEZETTEL;
revised by TOM PHILBIN

5 1/2 × 8 1/4 Hardcover 384 pp. 310 Illus.
ISBN: 0-672-23403-3 $14.95

WOODWORKING

Wood Furniture: Finishing, Refinishing, Repairing
(Third Edition)
JAMES E. BRUMBAUGH
revised by JOHN LEEKE

5 1/2 × 8 1/4 Hardcover 384 pp. 190 Illus.
ISBN: 0-02-517871-7 $25.00

A fully illustrated guide to repairing furniture and finishing and refinishing wood surfaces. Covers tools and supplies; types of wood; veneering; inlaying; repairing, restoring and stripping; wood preparation; and much more. Contains a new color insert on stains.

Woodworking and Cabinetmaking
F. RICHARD BOLLER

5 1/2 × 8 1/4 Hardcover 360 pp. 455 Illus.
ISBN: 0-02-512800-0 $18.95

Essential information on all aspects of working with wood. Step-by-step procedures for woodworking projects are accompanied by detailed drawings and photographs.

MAINTENANCE AND REPAIR

Building Maintenance
(Second Edition)
JULES ORAVETZ

5 1/2 × 8 1/4 Paperback 384 pp. 210 Illus.
ISBN: 0-672-23278-2 $11.95

Professional maintenance procedures used in office, educational, and commercial buildings. Covers painting and decorating; plumbing and pipe fitting; concrete and masonry; and much more.

Gardening, Landscaping and Grounds Maintenance
(Third Edition)
JULES ORAVETZ

5 1/2 × 8 1/4 Hardcover 424 pp. 340 Illus.
ISBN: 0-672-23417-3 $15.95

Maintaining lawns and gardens as well as industrial, municipal, and estate grounds.

Home Maintenance and Repair: Walls, Ceilings and Floors
GARY D. BRANSON

8 1/2 × 11 Paperback 80 pp. 80 Illus.
ISBN: 0-672-23281-2 $6.95

The do-it-yourselfer's guide to interior remodeling with professional results.

Painting and Decorating
REX MILLER and GLEN E. BAKER

5 1/2 × 8 1/4 Hardcover 464 pp. 325 Illus.
ISBN: 0-672-23405-x $18.95

A practical guide for painters, decorators, and homeowners to the most up-to-date materials and techniques in the field.

Tree Care (Second Edition)
JOHN M. HALLER

8 1/2 × 11 Paperback 224 pp. 305 Illus.
ISBN: 0-02-062870-6 $16.95

The standard in the field. A comprehensive guide for growers, nursery owners, foresters, landscapers, and homeowners to planting, nurturing, and protecting trees.

Upholstering
(Third Edition)
JAMES E. BRUMBAUGH

5 1/2 × 8 1/4 Hardcover 416 pp. 318 Illus.
ISBN: 0-02-517862-8 $25.00

The esentials of upholstering are fully explained and illustrated for the professional, the apprentice, and the hobbyist. Features a new color insert illustrating fabrics, a new chapter on embroidery, and an expanded cleaning section.

AUTOMOTIVE AND ENGINES

Diesel Engine Manual
(Fourth Edition)
PERRY O. BLACK;
revised by WILLIAM E. SCAHILL

5 1/2 × 8 1/4 Hardcover 512 pp. 255 Illus.
ISBN: 0-672-23371-1 $15.95

The principles, design, operation, and maintenance of today's diesel engines. All aspects of typical two- and four-cycle engines are thoroughly explained and illustrated by photographs, line drawings, and charts and tables.

Gas Engine Manual
(Third Edition)
EDWIN P. ANDERSON;
revised by CHARLES G. FACKLAM

5 1/2 × 8 1/4 Hardcover 424 pp. 225 Illus.
ISBN: 0-8161-1707-1 $12.95

How to operate, maintain, and repair gas engines of all types and sizes. All engine parts and step-by-step procedures are illustrated by photographs, diagrams, and troubleshooting charts.

Small Gasoline Engines
REX MILLER and
MARK RICHARD MILLER

5 1/2 × 8 1/4 Hardcover 640
ISBN: 0-672-23414-9 $1

Practical inform
maintain
en⸻ ⸻cycle
⸻ers, edgers,
⸻owers, emergency
e⸻ ⸻rs, outboard motors, and
ot⸻ ⸻ment with engines of up to ten
hor⸻power.

NEW EDITION FOR 1993

Truck Guide Library (3 Vols.)
JAMES E. BRUMBAUGH

5 1/2 × 8 1/4 Hardcover 2,144 pp. 1,715 Illus.
ISBN: 0-672-23392-4 $50.95

This three-volume set provides the most comprehensive, profusely illustrated collection of information available on truck operation and maintenance.

Volume 1: Engines
JAMES E. BRUMBAUGH

5 1/2 × 8 1/4 Hardcover 416 pp. 290 Illus.
ISBN: 0-672-23356-8 $16.95

Volume 2: Engine Auxiliary Systems
JAMES E. BRUMBAUGH

5 1/2 × 8 1/4 Hardcover 704 pp. 520 Illus.
ISBN: 0-672-23357-6 $16.95

Volume 3: Transmissions, Steering, and Brakes
JAMES E. BRUMBAUGH

5 1/2 × 8 1/4 Hardcover 1,024 pp. 905 Illus.
ISBN: 0-672-23406-8 $16.95

DRAFTING

Industrial Drafting
JOHN D. BIES

5 1/2 × 8 1/4 Hardcover 544 pp. Illus.
ISBN: 0-02-510610-4 $24.95

Professional-level introductory guide for practicing drafters, engineers, managers, and technical workers in all industries who use or prepare working drawings.

Answers on Blueprint Reading
(Fourth Edition)
ROLAND PALMQUIST;
revised by THOMAS J. MORRISEY

5 1/2 × 8 1/4 Hardcover 320 pp. 275 Illus.
ISBN: 0-8161-1704-7 $12.95

Understanding blueprints of machines and tools, electrical systems, and architecture. Question and answer format.

HOBBIES

Complete Course in Stained Glass
PEPE MENDEZ

8 1/2 × 11 Paperback 80 pp. 50 Illus.
ISBN: 0-672-23287-1 $8.95

The tools, materials, and techniques of the art of working with stained glass.

9 780020 777625